Uzman Khan
Muhammad Jahangir

Development of new analytical methods for toxicants used in textile

Uzman Khan
Muhammad Jahangir

Development of new analytical methods for toxicants used in textile

LAP LAMBERT Academic Publishing

Publisher:
LAP LAMBERT Academic Publishing
is a trademark of
Dodo Books Indian Ocean Ltd. and OmniScriptum S.R.L publishing group

120 High Road, East Finchley, London, N2 9ED, United Kingdom
Str. Armeneasca 28/1, office 1, Chisinau MD-2012, Republic of Moldova, Europe
Managing Directors: Ieva Konstantinova, Victoria Ursu
info@omniscriptum.com

Printed at: see last page
ISBN: 978-620-8-43389-5

CONTENTS

List of Figures

List of Tables

ISO methods and newly established method

Summary:

Different classes of synthetic chemicals are being applied to textile and leather related consumer articles. Most of them are regulated but they don't have certified testing methods that is why their presence cannot be measured on consumer articles. Therefore, new methods for testing of toxic substances were developed for leather, textile, foam, and plastic samples. Toxic substances were extracted with toluene, tetrahydrofuran, methanol, n-Hexane or ethyl acetate using a variety of experiments as an ultrasonic-assisted thermal extraction or water both with a shaker. Chemical toxicants were estimated using different techniques such as chromatography mass spectrometry (GC-MS) or ultra-fast liquid chromatography (UFLC). Pre-treatment limits were organized, including extractant type, method of extraction and time of extraction. Under prepared conditions, targeted compound retrieval was calculated. These proposed methods were successfully used for the determination of actual test specimens.

Recently regulated flame retardants, phthalates, biocide contents (MIT, CIT, OIT and Triclosan), Ortho-phenylphenols, DMFu and azo dyes (aromatic amines) don't have certified testing methods, therefore, their existence can't be measured in consumer products. Hence sensitive, reliable and fast methods were designed to classify and analyze the values of these controlled toxicants from textile, foam, leather and plastic samples. Subjected test specimens were extracted by selecting an appropriate organic solvent, temperature 60 °C, 60 min of sonication time and purified by utilizing centrifugation and filtration techniques. Toxic substances were examined either on GC / MS using a DB-5MS column and helium gas under Sim / Scan mode or on UFLC. Pre-treatment factors such as extraction solvent, extraction method, dilution ratio, extraction time were optimized. Under the optimized conditions, the targets had good linearity ($r^2 \geq 0.9995$). The findings were between $100 \pm 15\%$ for the subjected target analyte substances. The quantification limit (LOQs) was in the range of 0.02 - 0.2 mg L^{-1}. These proposed extraction and analytical methods were successfully utilized for the real experimental species containing many types of matrices. The toxic substances found were in the range as 5.18 mg L^{-1} - 1798.14 mg L^{-1}.

CHAPTER # 1

INTRODUCTION

1.1 Introduction

Analytical research is an important part of the textile industry. It plays an important role in the processing stages. This work is the essential part of laboratories in the textile industry. In addition, there is a need for quality assurance of textile products throughout their lives. This requires studying the interaction between the finished fabric product and the chemicals used during processing. Where there are residual solvents, wet and humid conditions, and other destructive reversals that may occur in the product for storage for some time under various stressful conditions. This includes the conditions under which they may be placed during storage or shipment of the finished products. The textile industry is increasingly being looked after by governments and civil society groups to keep costs down but continue to bring to market the safest and most efficient products that meet the needs of consumers. As part of the market-leading campaign, the industry has streamlined its operations in terms of textiles performance, and development. The way to bring new products to market quickly without compromising on quality or safety has prompted all companies to challenge all existing processes and seek ways to increase capacity, reduce deadlines, and do efficient using less resources. Analytical chemistry has played a main and key role for making the textile industry efficient. Indeed, the demand for analytical data has become a serious thing for the choice of target chemical agents for full development. Working under a subjected test specimen limited conditions and fully compliant with recent good manufacturing practices (cGMP), textile chemists are asked to maintain precise and accurate data. The technology used today and science, combined with the new binding regulations now, have made the analysis of the textile test specimens much more difficult than it was a decade ago.

The textile analysts play a major role in ensuring the efficiency, safety, quality and purity of the textile consumer products. The need for textile test specimen examination is largely driven by legal requirements. Legal considerations arise mainly when a commercial product does not meet its intended quality. New, flexible and self-employed processes in the textile processing industries are becoming more and more popular recently.

The number of new chemical agents used in the textile industry is increasing every year. Manufacturing industries cut off the use of regulated chemical agents and use alternative one. These chemical agents may be either new entities or partial structural modification of the existing one. In most cases, there is a gap between the time of entrance of a chemical agent in the market until the time of that chemical agent to addition in MRSL (manufacturing restricted substances list). This is the reason for the possible uncertainty in the ongoing and widespread use of these toxins. Complaints of new toxins lead to the withdrawal of those toxins from the market. Researchers are continuously working on development of long-suffering, nontoxic, eco & environment friendly, better chemical agents. These newly discovered chemical agents require standard analytical procedures.

Quality is the major concern in every product or service including textile and leather industry. The concept, quality control attempts to produce a complete product through a series of measures designed to prevent and eliminate errors in a different production phase. The decision to release or refuse a product depends on one or more types of regulatory actions. With the growth of the textile industry over the past decade, there has been rapid progress in the field of analytical chemistry involving the analysis and application of complex metals. Providing a simple analysis process for a complex matrix is a key point.

In short, the reasons for the development of new analytical methods were:

- A single chemical agent or a group of chemical agents need to be testing from a particular matrix but absence of validated methods.
- In the literature appropriate analytical method may not be existing for a particular chemical agent or matrix.
- Analytical methods may not be available for a particular chemical agent when that is present in the finished product matrix.
- Analytical methods for the quantitation of a single or a particular group of chemical agents may not be available they are present in the biological fluids.
- Analytical methods for a single or a particular group of chemical agents may not be available when they are present in the complex matrix.
- The existing analytical procedures may have complex sampling extraction, expensive reagents, less reliable and complex instrumentation.

Due to the progressing growth in the textile industry over the past decade, many types of matrix and new chemical agents have recently entered a major challenge, therefore, which is the reason for the rapid development in the field of textile examination. Improving the extraction and analysis methods of newly introduced textile chemicals is a top priority. Many modern types of analysis techniques are used to test the chemical agents used in the matrix. These analytical methods are required for standard adjustment and control testing. The most commonly used method for determining the quality parameters of textile-related articles is known as 'Chromatography'.

1.2 Chromatography

In 1906, Russian botanist M. Tswett reported the separation of various colored elements of green leaf extraction in a series of colored bands by allowing the solvent to flow into a column of calcium carbonate powder. He called this method 'chromatography' from Greek words meaning 'color (chroma)' and 'writing (graph)'. In fact, the truth of chromatography is based on his efforts. This approach did not appear in literature until the early 1940's, when Martin and Synge's famous paper was published. That paper reports the availability of liquid-liquid partition chromatography, columns and paper. They also presented a framework for the work of basic chromatographic processes and received the Nobel Prize in chemistry in 1952 against their work. The next big step that led to the advancement in this field was the invention of "gas-liquid chromatography" by James and Martin. The success of modern chromatography is due to the excellent management of chromatographic teaching by Giddings in 1965 with his book Dynamics of Chromatography. After that, many well-known scientists whose contributions are too numerous to recount here, and their work has led to the development of modern chromatography, commonly called high-pressure (HP) or high-performance liquid chromatography (HPLC).

The general classifications of chromatographic methods are as follows:

- Paper chromatography,
- Thin layer chromatography,
- Column chromatography,
- High performance liquid chromatography,
- Gel filtration chromatography,
- Chromatography of Supercritical fluid.
- Ion-exchange chromatography,
- Gas chromatography

1.3 High performance liquid chromatography:

High performance liquid chromatography performance (HPLC) is a modern method of analysis. That analytical technique is employed in analytical chemistry to identify, classify and measure component of mixtures under examination. In this technique pump is used to pass liquid solvent that contain the test specimen mixture that pass through a column packed with solid adsorbent material. The components of the test specimen interact differently with adsorbent objects, resulting in different flow rates of test specimen objects. That leads to the separation of analytes as they come out of the column. The HPLC instrument typically includes pumps, degasser, sampler, and detector.

1.3.1 Operation:

- The test specimen mixture is presented in small volumes (microliters) in the flow stream of mobile phase that passes through the column. Parts of the test specimen pass through a column at different speeds. The velocity depends on the nature of the stationary phase, the structure of the moving phase and the chemical nature of the testing material. The whole time period taken for a specific analyte to elute is termed as the retention time. Retention time is a specific element of a given analyte.

Different types of columns are available on the market, packed with adsorbents that vary in type of chemical and particle size. Small particle size packaging materials require high operating pressure and develop a high degree of separation between successive analytes that were analyzed by a particular column. Sorbent particles can be hydrophilic or hydrophobic naturally, but it depends on choice and requirements of the analytes.

Mobile phases that are used commonly contain any miscible combination of water with numerous organic solvents. Acetonitrile and methanol are frequently used. Some HPLC techniques use water-free mobile phases. In that techniques the aqueous component of the mobile phase contain acids as phosphoric, formic or trifluoroacetic acid or salts of acids to assist the separation of the test specimen components. The composition of the mobile phase may be kept constant, isocratic elution or varied, gradient elution during the chromatographic analysis. Typically, isocratic elution is more effective for the separation of test specimen components that are different in their affinity for the stationary phase. In gradient elution the composition of the mobile phase is varied mostly from low to high eluting strength. The eluting strength of the mobile phase is reflected by analyte retention times with high eluting strength producing fast elution of the components of the mixture. Reversed phase chromatography, in gradient elution mode may have multiple choices as, it can be start at 5 % acetonitrile 95 % water or liquid buffer and proceed linearly to 95% acetonitrile 5 % water or liquid buffer within 5–25 minutes. Periods of constant mobile phase composition are a specific part of any gradient elution profile. Mobile phase composition may be kept constant at 5% acetonitrile in water or aqueous buffer within 1-3 min and can be changed a linearly as, 95% acetonitrile in water or liquid buffer.

1.3.2 History:

Previous to HPLC scientists used standard chromatographic liquid methods. Water chromatographic schemes were ineffective due to the flow rate of gravity-dependent solutions. Separation takes long time as sometime it takes hours or days. Gas chromatography (GC) was more controlling at that era than liquid chromatography (LC). However, it is general thinking that the analysis of gas phase test specimens and molecular analysis of highly polar biopolymers were extremely impossible at that time. The Gas chromatography is not effective for many biologists due to thermally unstable solute particles. So, here need for alternatives were considered that's resulted to develop HPLC.

Succeeding the work of Synge's corpses and Martin in 1941, Josef Huber, Cal Giddings, and some other scientist predicted in the 1960 that liquid chromatography could be functioned proficiently by reducing the range of packaging particles under normal LC (and GC) by 150 μm and using pressure to increase mobile phase speed. These predictions were widely tested and refined between the 60s and 70s. Early developmental research began to improve LC particles, and the construction of Zipax, promised HPLC technique.

Hardware and instrumentation of HPLC become more advance in 1970s. Researchers just start to use injectors and pumps to achieve the elementary design of the HPLC system. Gas boost pumps were perfect the reason was; they can be operated at continuous pressure, not require stagnant signals or valves to check constant flow and good volume. Important computer steps were performed in the Industrial Polymers Division as a low-grade gradient tool used with the installation of a loop injection valve and septum injector.

1.3.3 Types:

1.3.3.1 Partition chromatography:

This type of chromatography was one of the first forms of chemical separation chromatography. The equilibrium-equilibrium principle has been employed in the paper chromatography technique. It's were in thin layer chromatography, gas phase and liquid-liquid separation techniques. Nobel prize in chemistry dated 1952 was presented to Richard Laurence Millington Synge and Archer John Porter Martin for the development of partition chromatography, that was utilized first time for amino acid separation. This chromatography uses a stored solvent, on the surface or inside the breasts or strips of solid "inert" matter like paper chromatography; or take advantage of a combination of donor coulombic and / or hydrogen and stable phase. Separate the molecules under examination between the standing water phase and the eluent. That process is just similar to hydrophilic interaction chromatography (HIC). That technique separates the analytes according to their polarity differences. HIC is most commonly used with a polar-bound molecule and a mobile section made primarily of aqueous acetonitrile as a strong component. Separation HPLC has historically been used in unmixed silica or alumina foundations. Each is effective in classifying analyzes by polar-related variants HIC boundary segments have the potential to separate neutral, acidic or basic solutions into just a single run.

1.3.3.2 General phase chromatography:

Normal phase chromatography was one of the first types of HPLC that chemists developed. Also known as normal phase HPLC (NP-HPLC). This method differentiates analytes based on their relative affinity with silica and the analyte ability to participate in polar connections (as in dipole-dipole type or hydrogen-bonding) on the surface of the sorbent. NP-HPLC uses a liquid phase, a waterless phase (e.g., Chloroform), and is effective for examining of analytes that are easily soluble in non-polar solvents. The probe converts and retains a stable polar phase. Adsorption power increases with increasing analyte polarity. Main separation power works on the functional groups attached on the structure of the analyte molecule, combined with some specific factors. The effect of a strong barrier to co-operation allows this method of differentiating isomers of the same structure.

1.3.3.3 Displacement chromatography:

The basic principle of displacement chromatography is: A molecule with a high degree of chromatography matrix (displacer) will compete effectively in binding sites, and thus remove all molecules with small affinities. There is a distinct difference between migration and elution chromatography. In elution mode, objects usually appear in small columns, Gaussian peaks. Extensive separation of the basic analyte peaks, is necessary to achieve the highest purity. The composite molecules elute in the column depends on many factors. For two substance molecules to move at dissimilar speeds, and thus be resolute, there must be a substantial difference in the interaction between the subjected substance molecules and the chromatography matrix. Performance limitations are accustomed to enhance the effectiveness of this dissimilarity. In most of the cases, the basic peaks separation can only be attained by a combination of gradient and loading of the lower columns. Therefore, two issues in the elution chromatography, especially on the preparation rule, are complex in operation, due to solvent gradient pushing, low flow, and the low loading of the columns. This chromatography is more advantageous than elution chromatography in those objects that are solved into successive parts of pure material rather than "peaks". Because the process utilizes isotherms inconsistencies, large column feeds can be separated from the column provided by the refined material obtained at the highest concentration.

1.3.3.4 Reversed-phase chromatography (RPC):

Reversed-phase chromatography has a non-polar phase and a slow, medium-phase water phase. One common standing class is silica converted further by RMe_2SiCl, where R is a direct alkyl group such as $C_{18}H_{37}$ or C_8H_{17}. In these vertical phases, the end time is longer in the lower extremities, while the polar molecules move more easily (at the beginning of the analysis). The detector can extend the end times by adding more water to the mobile phase section; therefore, turning the hydrophobic process closer to the hydrophobic phase stronger compared to the slow-moving cell phase. Similarly, the detector can reduce the final time by adding more organic solvent to the eluent. RP-HPLC is so widely used that it is often mispronounced as "HPLC" without further clarification. The textile industry regularly uses RP-HPLC to obtain drugs before they are released.

RP-HPLC works on the principle of hydrophobic interactions, which range from high concentrations in the structure of the dipolar water and play a very important role in all processes. The RP-HPLC allows for the measurement of these cooperative forces. The binding of the attribute in the vertical phase is equal to the contact area around the non-polar part of the analysis molecule when it connects to the ligand in the vertical phase. This solvent repellent effect is dominated by the water “reduction hole” around the analyte and the C18-chain compared to your dual complexity. The energy extracted from this process is equal to the surface strength of the eluent and the hydrophobic surface of the analyte and ligand respectively. Storage can be reduced by adding only solvent (methanol, acetonitrile) in the mobile unit to reduce water resistance. The gradient correction utilizes this effect by automatically reducing the interaction and the surface height of the water phase leading to the analysis.

1.3.3.5 Size-exclusion chromatography:

Extrusion chromatography (SEC), also known as gel filtration chromatography / gel permeation chromatography, classifies particles based on particle size. Chromatography has a low resolution and is therefore subject to a final, "polishing" step of cleaning. It is also helpful in determining the quaternary structure of refined proteins. The SEC is mainly used for the analysis of large molecules such as proteins or polymers. The SEC works by trapping these tiny molecules in the holes of particles. Larger molecules pass through the pores as they are too large to penetrate the pores. Larger molecules therefore flow in the column faster than smaller molecules, i.e., when the molecule is small, the longer it is stored.

This method is widely used in determining the molecular weight of polysaccharides. The SEC is the official procedure for comparing the weight of low-weight heparin cells obtained commercially.

1.3.3.6 Ion-exchange (IC):

Separation of the components of the mixture is based on the interaction between solute ion particles and charged sites bounded to a fixed phase in ion-exchange chromatography. In ion-exchange chromatography the solute ion particles of the similar charge and the charged site mites inside the column are detached from the binding sites, while the less charged ion species on the column sites are stored inside the column material. Column ions stored in the column can be detached from the column material sites by altering the solvent behavior. That process would be in action by just enhancing the ionic effect of the solvent system used as by increasing the concentration of salt in solution, temperature elevation of the column oven, altering the pH value of the solvent, etc.

1.3.3.7 Bio-affinity chromatography

This chromatographic process relies on the stable, precise and flexible structures. The formation of these complexes involves the participation of similar molecular forces such as Van der Waals interactions, electrostatic interactions, dipole-dipole interactions, hydrophobic interactions, and hydrogen bonds. An efficient, bio specific bond is composed of the simultaneous and combined action of most of these forces in the corresponding bindings.

1.3.3.8 Normal-phase chromatography:

That chromatographic technique is the combination of a mobile reversed phase chromatography and the standard organic phase chromatography. This technique is used to examine a variety of hydrophilic chemicals agents. It works as normal chromatographic technique but it utilizes reverse phase elution process.

1.3.4 Types of elution:

The elution process in which the structure of the mobile phase remains unchanged in the whole elution process is termed as isocratic elution. An example of this is the %age of acetonitrile in an examination will remain unchanged at 20% by volume. The name "isocratic" was first time sued by Csaba Horvath, he was one of them the HPLC pioneers.

Mobile phase design may not be always constant. When the structure of the mobile phase is altered during the elution process that elution process is defined as the gradient elution adjustment. Here an example of a gradient process is, an elution process that starts at 20% acetonitrile go on and terminate the elution process at 80% acetonitrile after 15 minutes. The parts of the mobile phase section are called component "A" & component "B". Solvent A is usually a weak component that allows the solute particles to move slowly, on the other hand component "B" is usually a strong solution that moves the solute particles rapidly through the column. In the gradient chromatography phase, Component "A" is usually an aqueous buffer and "B" may be a water-soluble solution or water based organic solution as methanol, THF, acetonitrile or isopropanol.

1.3.5 Parameters:

1.3.5.1 HPLC examination has some theoretical parameters for defining the separation of the components of a mixture the form of signal peak detected by instrumentation such as by mass spectrometer or UV detector. Theoretical parameters are found mostly in 2 sets of chromatographic theories, one of them is Van Deemter equation and second one is plate theory (its particularly for partition chromatography). These two theories can be used commonly for the examination of HPLC chromatograms, it is general thinking that theoretical measure is a precise examination theory.

1.3.5.2 Void volume:

It is the empty space that remain in the column when it is filled with the mobile phase solvent. It is the empty space inside the column and the internal packaging of the column material. The exact idea of empty volume can be taken from a chromatogram. From the start of the chromatogram to the first high value peak is appeared. Most commonly the solvent existing in the test specimen components. Infect, the solvent in the test specimens flows through the column but it does not contact with the column packing material. That solvent shows different behavior in the chromatogram and seems to be different from that of the HPLC solvent. Empty volume is used as a correction factor.

1.3.5.3 Efficiency factor (N):

It measures the sharpness of the chromatogram peaks, as the fractional area ("end time"). It belongs to the wider area of the peaks (peak base). Longer, sharper, and lower altitudes of peaks indicate that the separation has effectively eliminate some of the components. Efficiency depends largely on the HPLC column and the HPLC method used. Efficiency is the same as number of theoretical plates or plate number.

1.3.5.4 Retention factor (kappa prime):

It gives the estimation of the amount of component attached to a column. It measures the area below the height curve in its chromatogram (because HPLC chromatograms are a time function). Each chromatogram value will have its own final feature (e.g., Kappa-1 for the first high-end component). This feature can be configured with an empty column volume.

1.3.5.5 Separation factor (alpha):

It is the comparison associated with how two adjacent parts of a chromatogram are separated. This item is defined by a pair of high-resolution neighboring chromatograms and can be adjusted to an empty column volume. When the value of the separation factor is greater than 1.0, it is better to divide, up to about 2.0 when the HPLC method may not be required separately. The separation factor makes the relation among three factors efficiency and classification of factors that improves the peaks resolution in HPLC.

1.3.5.6 The internal diameter of the column:

HPLC column is a significant thing that impacts on the selectivity and sensitivity for gradient elution components. Internal diameter is the main factor that controls the volume of the analyte which could be filled in the column space. Industries use larger/longer columns for the purification of drug products. That practice can be seen specifically in pharmaceutical industries. Columns that have short internal diameter have advantage to consume less solvent and their sensitivity response is better. Columns with larger internal diameter that may be over 10 mm are in common practice to filter the material values due to their large loading capacity.

1.3.5.7 Particle size:

Common HPLC contain a fixed component attached, except for small very small circular beads. These circular beads are available in variety of sizes in the market out of which 5 beads size are used commonly. Small size beads have the advantage to provide more space, better separation of components so good resolution, but columns with smaller size particles have more pressure than that of bigger particle size columns.

1.3.5.8 Pore size:

To attain more surface area, stationary phases having modifications as some have porous inside. Smaller pores provide more space on the other hand larger pore size having advantage for better movement of solute/solvent particles, especially for analytes having larger particle size. It can be better considered from a protein molecule that is larger in size. That protein molecule is just slightly smaller than that of the pore size of the column material. It would move inside the pore but due to having larger size its movement would be slow inside the column. The effect of temperature, time, concentration of mobile phase and solid to liquid ratio would work less on that type of analyte having larger particle size.

1.3.5.9 Pump pressure:

Pumps vary in compression capacity, but their performance is measured by their ability to produce a constant and volatile volumetric flow rate. The pressure can reach as high as 60 MPa (6000 lbf / in2), or up to 600 atmospheres. Modern HPLC systems have been optimized to operate at very high pressures, so they are able to use very small particle sizes in columns (<2 μm). These "ultra-high-performance liquid chromatography" or UHPLC systems can operate up to 120 MPa (17,405 lbf) or approximately 1200 atmospheres.

1.3.6 Detectors:

HPLC usually uses UV-Vis detector but a variety of other chromatography detectors can also be used in liquid chromatography. A universal detector that accompaniments UV-Vis detection is an aerosol detector. The most widely used detectors include refractive index, that detector show reading by assessing ions in the eluant indicator as they flow. Combined utilization of different types of detectors at the same time in an instrument is also possible in some cases. That can be seen in LCMS instrument which use UV-Vis detector combined with the mass spectrometer.

1.3.7 HPLC Autosampler:

A variety of autosamplers are available in the market commercially. A lot of test specimens can be loaded to the HPLC system automatically. That action can be performed with the help of autosamplers attached with the instrument. Autosamplers have injectable proportions and technologies that provide constant injection volume.

1.3.8 Applications:

1.3.8.1 Production:

There are many applications of HPLC instrument. Vast use of that instrument can be seen in industries, particularly in laboratories and medical science examinations. To ensure the purity, quality and quantity of the product components that instrument is commonly employed in drug manufacturing, textile and chemical processing industries.

1.3.8.2 Legal:

This method is used to detect prohibited drugs from the urine test specimens. Immunoassays is the frequently used method for drug discovery.

1.3.8.3 Analytical research:

To obtain authentic and constant results HPLC is frequently utilized in analytical research and to find out the goals of people seeking treatment such as anti-fungal drugs and asthma etc.

1.3.8.4 Medical:

In the medical field HPLC is commonly used for the analysis of illicit drugs. Mostly the urine test specimens are examined to find out the concentration of a particular drug. Blood test specimens for blood serum levels designed for clinical analysis are also analyzed by HPLC.

1.4 Gas chromatography (GC):

It is a common type of analytical process that is most commonly used for the separation, quantitative and qualitative analysis of chemicals. That analytical technique is utilized for the separation of various elements of a component or testing the purity of a subjected compound. Exact values or estimated values of a substance can also be found. GC may assist in computer identification of substance values. That analytical technique can help to separate pure components from the compound in preparative chromatography.

The mobile phase is a gas most commonly helium or nitrogen gas. Helium remains the main energy source used by about 90%, though hydrogen is preferred for advanced separation. Stationary phase is a thin layer of liquid or polymer on a solid inner support, inside a piece of glass or a metal tube called a column. The instrument used to perform gas chromatography is called a gas chromatograph (aerograph or gas separator).

1.4.1 History:

Chromatography came in 1903 from the work of Russian scientist Mikhail Semenovich Tswett, who distinguished plant colors by a chromatography column. Austrian graduate student Fritz and German chemist Erika Cremer

(1947) first time make the way for the foundations for GC ideas and made the first gas-liquid chromatographic technique. This work was considered insignificant and long overlooked. Archer John Porter Martin awarded the Nobel Prize for his work on developing liquid - liquid chromatography in 1941 and paper chromatography in 1944, so it is said to be the basis for gas chromatographic technique. The development of flame ionization detector (FID) laid the dramatic change in the popularity of gas chromatographic technique.

1.4.2 Analysis at Gas chromatographic technique:

Gas chromatographic technique is a common analysis tool for the chemical analysis of complex test specimens. A flow tube is used also known as a column; the components of the test specimen pass through that column. Carrier gas moves these components. Components of the test specimen moves at dissimilar values dependent on their properties, chemical composition and their complementation by a particular column. As the components of the mixture come out of the end of the column, these are separated so, can be analyzed by the detector. The column is most commonly in vertical position to enhance the separation of the components of the mixture. All the components move freely but their speed is not same so, each of them come out of the column at their own specific time. That time is referred as retention time. Column length, gas flow rate and temperature can be varied to change the order of that end time.

A specific amount of gaseous test specimen or analyte in the form of liquid is injected into the column, most commonly with the help of a small syringe, gas source switch system or microextraction fiber strands. Carrier gas propels the test specimen molecules into the column, this flow prevents the stuck off the test specimen components in the column walls and also reduce the time for attraction with the column packaging material. The extent to which molecules progress in a column depends on the potential of the adsorption, which also depends on the type of molecule and the phase properties. Since each type of molecule has a different level of continuity, different parts of the analyte compound are separated as they progress in column and reach the end of the column at different times (retention time). The detector is used to monitor the stream from the column; therefore, the time at which each component reaches the outlet and the value of the item can be determined. Typically, items are identified (to some extent) in their (elute) sequence from column and time of analyst in the column.

1.4.3 Physical components:

1.4.3.1 Autosamplers for Gas chromatographic technique:

Test specimen can be introduced automatically in the input areas with the help of autosamplers. Test specimen can be injected by different ways. Better reproducibility and injection time frame can be achieved by installing an autosampler.

Autosamplers are available in the market in the form of a multiple type of varieties. They can be classified with respect to the strength of the test specimens they can operate and inject. Auto-injectors can operate and inject a small and specific number of test specimens. Autosamplers can operate a vast number of test specimens in a variety of ways.

1.4.3.2 Inlets;

S / SL injection (split / splitless):

The test sample is placed in a small warm room with a syringe through the septum - heat helps the test specimen and matrix to volatile easily. Test specimen in the volatile from sweep in the column with flowing carrier gas. Either a part of the test specimen / carrier gas mixture is injected in the column with the help of an injection split vent (split injection system) or a whole portion of the test specimen / carrier gas mixture is injected in the column with the help of an injection (splitless injection system). When working with such test specimens having high concentrations of analyte in the test specimen (> 0.1%) in that situation split injection system is best suited. For test specimens having low concentrations of analyte (<0.01%) splitless injection system is better. The separating valve opens after a specific pre-set time to eliminate the entry of heavy objects that could disturb or contaminate the system in split injection. The shorter duration pre-set time in split injection system (< 0.2 min) ensures less peak tailing but on the other side loss in response may occur. Longer duration of pre-set time (>2 minutes) may enhance the peak tailing factor but better signal response is received.

1.4.3.3 On-column inlet:

The test specimen here is delivered directly to the column completely without temperature treatment or the provided temperature is below the boiling point of the solvent in On-column inlet. Low temperatures complement the test sample in a small area. Column and entry can be heated test specimen vaporizes and covert into gas phase. This ensures a much lower temperature for chromatography and make possible for test specimens to disintegrate beyond their boiling point.

1.4.3.4 PTV injector:

The temperature for the experimental sample was first described by Vogt in 1979. Vogt originally developed this method as a way to introduce large volumes up to 250 μL for capillary Gas chromatography. Test specimen was injected by Vogt into the liner with constant injection rate. Liner temperature is selected just a little less than the boiling point of the selected solvent. Heated solvent evaporates frequently and come to the separating line through vent valve. According to this procedure, Poy established a constant heat injection (PTV). So, introducing a test specimen at low temperature liner, the magnitude of the problems of old heat injection techniques can be reduced.

1.4.3.5 Gas source or gas exchange port:

Various types of gas phase test specimens in the assortment bottles are coupled to what is usually a 6-port valve switch. The carrier's gas flow is not affected when the test specimen extended into a previously removed test specimen loop. Loop are incorporated into the carrier gas stream, when switching the contents of the test specimen.

1.4.3.6 Purge & trap program:

Flowing electricity is struck by a liquid test specimen that causes the purification of volatile chemicals in the matrix. At ambient temperature, volatile liquids are trapped in a sorbent column. Heating up the trap transfer the volatile liquids directly into the flowing carrying gas. Test specimens that require cleaning are imported through such a process, mostly connected to the purge & trap port.

1.4.3.7 Carrier gas selection:

Hydrogen has better flow rate compared to other gasses just like helium. If flow levels are properly maintained, helium may be more efficient and provide better separation. A large number of old machines and equipment operate on helium as helium is non-flammable. It is general thinking that helium is the most widely used human gas. Recently most chromatographers switch to hydrogen gas as the price of helium has risen sharply because of its high consumption and demand. With reasonable consideration of its history, in the future helium may be continuously used in instrument.

1.4.4 Detectors used in the gas chromatography:

Thermal conductivity detector (TCD) and the flame ionization detector (FID) are the most widely used devices. Both of these work with different concentrations and are sensitive to different types of components. Except the gas-carrying used to propel test specimen TCDs detectors can analyze any type of test specimen. Due to their versatile nature, they are considered universal. FIDs are unable to detect water from test specimens. FID detectors are more sensitive than TCD. They can detect hydrocarbons more efficiently than TCD. In case of sensitivity FIDs are more sensitive and efficient than TCD. FIDs and TCDs both detectors are very sensitive and robust. FID and TCD both can be used in series as FID is destructive but TCD is non-destructive so, in series they can work in combined form. They would provide reliable results for similar analytes. Most of the available detectors either can work in narrow range of concentrations of analytes or they are sensitive to certain specific types of test specimens only.

1.4.4.1 Thermal conductivity sensor (TCD):

These detectors are dependent on the thermal conductivity of the rhenium & tungsten fiber through which the current can flow. In this preparation helium or nitrogen acts as a gas carrier due to its high thermal conductivity which keeps the filament wire at low temperature so, maintains the same resistance and electrical efficiency. From the column the analyte molecules when emerges they mix with the carrier gas, due to that phenomenon the thermal conductivity to detector decreases and this creates a detector response problem. The answer is due to the decrease in heat flow leading to increased filament temperature and resistivity leading to a change in electrical energy. The sensitivity of the sensor is similar to that of the present and differs greatly from the low temperature and humidity of the detector.

1.4.4.2 Flame ionization detector (FID):

In these machines' electrodes are placed near a hydrogen flame near the outlet of the column, where carbon-containing compounds emerges from the column they are pyrolyzed by the flame. This detector uses only organic / hydrocarbons containing chemicals due to the ability of carbon to form cations and electrons in addition to the pyrolysis that produces what is present between the electrodes. The current increase is interpreted and seen as chromatogram height. Flame ionization detectors show less visual acuity just a few picograms / second and cannot produce ions from carbonyl-containing carbon. Hydrogen, nitrogen, helium and argon are used as carrier gasses in flame ionization detectors

1.4.4.3 Alkali flame ionization detector (AFID) or Alkali flame detector (AFD):

These detectors have high sensitivity to phosphorus and nitrogen, just as nitrogen-phosphorus detector. Instead of beads on flames, the alkaline ions are provided with hydrogen gas. AFD provides continuous sensitivity over long period of time and not experience the fatigue as NPD. AFD acts just as a normal flame ionization detector when alkali ions are exposed to flames. Combustible hydrogen & hydrocarbons can be examined using catalytic combustion detector (CCD). To produce ions, discharge ionization detector (DID) uses high power voltage.

1.4.4.4 The poly-arc reaction detector:

It is the addition in GC-FID. All naturally occurring molecules are breakdown to methane than they are detected by ionization detector. This method can be used to improve the FID response. It makes possible the analysis of many carbon-containing chemical compounds. Because the reaction characteristics are similar to that of methane so, here no need for reference standard but for this all the chemical compounds are converted into methane. Poly-arc reaction makes possible the easy analysis of even complex chemical molecules and compounds without reference standards.

1.4.4.5 Flame-photometric (FP) detector:

It uses ignited flame and photomultiplier tube to examine spectral line. Certain elemental wavelengths of exited molecules (S, P, Halogen or other metallic elements) are used to analyze chemical compounds that are burned in hydrogen-based flame on exiting column outlet. A photomultiplier tube detects the emitted and filtered light. Sulfur emission is at 394 nm and phosphorus is approximately 510-536 nm. A powerful microwave compartment is used that generates plasma in atomic emission detector (AED). The subjected test specimen eluting from column outlet is treated with microwave and plasma. Test specimen decomposed on treatment to plasma. Specific elements produce atomic emissions spectrum when the test specimen is treated with plasma. An image diode or series of photomultiplier tubes are used to detect atomic emission spectrum and diffraction grating is used to separate spectrum lines.

1.4.4.6 Electron scanner (ECD) detectors:

To measure electron capture images ECD detectors uses a beta source of radiation "a beam of electrons". Carbonyl, nitriles, halogen, organometallics and nitro groups like active groups or molecules are detected by ECD detectors. In ECD detectors argon plus 5% methane or nitrogen gas is used as a control gas. Radioactive substance like Ni-63 is placed in front of cathode electrode and two electrodes with active current are placed at the end of the column. A beam of electrons (beta particles) is generated by radioactive foil that drops and ionizing the gas carrier to produce more lead ions leading to the current. When molecules under examination have electron emission components or active groups of electrons are taken, they lead to a decrease in the current output of the detector response.

1.4.4.7 NPD detectors:

phosphorus and nitrogen transform into specially dressed beads and the current effect is measured in nitrogen phosphorus detectors. NPD detectors are special types of thermionic detectors.

1.4.4.8 Dry Electrolytic Detector (DED):

High temperature and atmospheric air were used to measure the chlorine based chemical compounds in DED detectors.

1.4.4.9 Mass spectrometric detector (MS):

GC-MS abbreviation is used for gas chromatographic mass spectrometric detectors. These detectors are very sensitive, perform very efficiently and very small amount of test specimen is required. This detector can be used to target analytes in the form of chromatograms in bulk. NMR display can be connected to a gas chromatographic mass spectrometric detector that can be utilized as a backup detector. GC/MS/NMR abbreviation is used for that setup. Infrared (IR) spectrophotometer can also be used with GC/MS/NMRs setup here IR spectrophotometer would act as a backup detector. GC/MS/NMR/IR abbreviation is used for that setup. Most of the required analysis can be conducted with GC-MS only due to this reason GC/MS/NMR/IR setups are very rare in practice.

1.4.4.10 Vacuum ultraviolet detectors (VUV):

In gas chromatography machines vacuum ultraviolet is the most recent advancement. Many types of chemicals absorb and have a wide range of gas phase absorbance at vacuum ultraviolet wavelength range of 120-240 nm approximately. In the absence of chemical interference, when the cross-absorption components are known analytes, vacuum ultraviolet detectors can determine the unlimited number of molecules present in a flow cell.

Detectors are also present other than presented above these are helium ionization detector (HID), Hall electrolytic conductivity detector (HECD), thermionic ionization detector (TID), infrared detector (IRD), pulsed discharge ionization detector (PDD) and photo-ionization detector (PID).

1.4.5 Conversion of test specimen into gas phase:

The subjected test specimen in gas chromatography is converted into gas phase for examination by using two valves. Injector loop is charged with test specimen in the form of gas, both valves are opened, pressure is applied in the form of carrier gas to the test specimen loop and the test specimen is pushed by force into the column for the separation of test specimen components.

The collection of required necessary parameters for the operation of the gas chromatographic instrument for a given analysis is termed as method. The procedure of defining which specific parameters may be sufficient or appropriate for the subjected analysis is termed as method development.

Test specimen input rate, temperature of the column, temperature of the detector, temperature of the carrier gas, temperature of the operating system, standard column width, length, type of flow of carrier gas, flow rates of carrier gas, test specimen size and injection process are the conditions that can be changed for a specific analysis. Depending on the equipment installed in the gas chromatography, there may be many acquisition conditions that may also be varied. Some gas chromatography equipment's incorporate valves that can change the flow rate of carrier gas and flow rate of the test specimen injection method. That is done by controlling the timing of the opening and closing of these valves.

1.4.6 Gas selection and gas flow rates:

Helium, nitrogen, argon, hydrogen and air are the commonly used carrier gasses. The choice of the carrier gas is mostly depending on the detector used, e.g. for analyzing gas test specimens on discharge ionization detector carrier gas helium must be required. Sometimes a specific carrier gas may be selected according to a specific test specimen matrix, e.g. argon as a carrier gas is preferred for test specimens that are mixture in argon gas, the reason is that on the chromatogram argon peak does not appear. Carrier gas selection may depend on availability or safety e.g. hydrogen burns and high-purity helium can be difficult to find in other parts of the world. In many applications and in most part of the world, due to the shortage of helium, hydrogen is used as a carrier gas instead of helium.

The detector used determines the purity of the selected carrier gas, the required sensitivity level of a carrier gas may also play a major role. Generally, 99.995% pure carrier gas is required. Modern instrument requirements are stricter as they work on highly pure grades of carrier gas 99.999% purity (>5.0 grades). Higher grade standards are often used as, 6.0 grade, but the need to reach very low levels of other environmental and legal applications has necessitated a 7.0 grade cleanliness and is now available for sale. Ultra - high purity range (UHP), Zero range, 4.5 range and 5.0 range these are some common grades of purity range that are commercially available in the market.

Such like temperature the flow of carrier gas affects the analysis in the similar manner. If the speed of the carrier gas will be high, the analysis will be quicker and, in less time, but the components of the mixture would be less separated hence results will be less reliable. So just like column temperature selection, the velocity or flow rate of the carrier gas is much important. The selection of appropriate flow rate will affect directly the separation of the components of the mixture and analysis time. Line velocity can be calculated using the gas flow rate, relative to the column width.

By controlling the input pressure, the flow rate of the carrier gas and the pressure of the column head space was adjusted indirectly in gas chromatographic instruments developed before the 1990s. The actual flow rate measurement was time-consuming and problematic process in gas chromatographic instruments developed before the 1990s. A bubble flow meter or an electronic flow meter was used to measure actual flow rate of the gas chromatographic instruments. A bubble flow meter or an electronic flow meter was attached at the outlet of the detector of column. During the analysis varying the flow rate was not possible in gas chromatographic instruments developed before the 1990s because the flow rate changing during analysis make the flow rate pressure non constant hence less reliable results. For compressible fluids the Poiseuille's equation is the best way to measure the relationship between inlet pressure and flow rate.

The pressure of the carrier gas is electronically controlled and the flow rate of the carrier gas is estimated by computer aided programs in most of the modern gas chromatographic equipment's. Hence similar to temperature programs, flow programs /pressure and adjustment of carrier gas flow rate and pressure during analysis is possible in that modern gas chromatographic equipment's.

1.4.7 Selection of the stationary phase components:

The stationary phase compound selection is most important for analysis. The stationary phase compound should have the same polarity as the solute so, polarity of solute is most important for the stationary phase compound selection. In open tubular columns the stationary phases used are carbowax polyethyleneglycol, biscyanopropyl, diphenyl dimethyl polysiloxane, cyanopropylphenyl polysiloxane, and cyanopropylphenyl dimethyl polysiloxane. In the commercial market, columns are available with a variety of available packing categories.

1.4.7 Flow rates & column inlet varieties:

The subjected test specimen may be in gas, liquid, solid form or adsorbed onto a substrate. So, the choice of the injection process or column inlet type may depend on the test specimen type. Soluble The test specimens that are easily vaporizable and soluble in complete form can be inserted directly into the column with the aid of the injector (COC) in that situation the analyzing parameters should be well known. In S / SL injection system (standard injection procedure) matrix is slightly removed with glass wool here gas test models are installed using a gas exchange system as cylinders filled with air. In case of the test specimens that are adsorbed onto a substrate as for test specimen that is captured in an adsorbent tube these test specimens are inserted with the help of an external (off-line/inline) extraction device such as a cleaning and holding device or inserted into an injector (SPME systems).

1.4.9 Test specimen injection in gas chromatography:

1.4.9.1 The tenth (1/10) rule:

Just after the test specimen injection in column head space the gas chromatographic analysis begins. Prier steps are test specimen processing and that are not part of analysis. In gas chromatography real examination of test specimen start after injection of test specimen in the column. Scientists face many problems associated with the injection process especially in capillary gas chromatography. The reason is that the capillary columns are different from other columns and the injection procedure used for other types of columns are not possible with capillary columns. As the chromatographic process is different in capillary gas chromatography so the width of the inserted plug should be small compared to the spread and the inserted volume should not overload the column. The power of the column separation will be reduced if column will be overloaded with test specimen. The V-inj (injected volume), / the V-det (detector volume), that resides of the test specimen after passing the test specimen from the column should be about 1/10 of its original volume for capillary gas chromatographic analysis.

For efficiently and suitable separation in columns the general requirements for a suitable injection procedure may be, it may not cause a change in the structure of the test specimens, it may allow complete and accurate injections of small amounts of test specimens to represent, it may be used for untested test types and track analysis and it may not indicate a difference in boiling points, concentration, polarity or thermal stability.

Other than these stated above, the injection systems have number of general problems. In the wrong hands or when the operator is unskilled, errors are very serious and non-controlled, even the best injectors claimed by their manufacturer have only accuracy level up to 3% only. Every time when needle is inserted into septum, it cut small pieces of septum rubber. These pieces accumulate in the syringe, can block the filling system of the syringe and next time one may fail to refill the syringe. That problems are general and may or may not occur frequently. Sometimes one can face a problem that the chromatogram shows ghost rise in the peaks that is the reason of a part of the test specimen stuck from previous injection and release in the subsequent injection. In other case a part of the test specimen material components may evaporate from the end tip of the syringe needle and the remaining part of the test specimen may show elevated results.

1.4.10 Selection of the column for a particular test specimen:

Applicable scale and the test specimen components play a major role for column selection. The test specimen internal chemistry as the polarity of the components of the mixture and the functional groups attached can lead towards a specific direction for the selection of a particular column. To reduce run time for analysis and to increase column resolution, the polarity of the column stationary phase and the polarity of the test specimen components must closely match with each other. Test specimen total run time and separation of the test specimen components mainly depend on the release of the film (vertical section), column width and column length.

1.4.11 Temperature scheme:

Columns in GC are contained in the oven, the temperature of which is carefully controlled by a computer. When referring to "column temperatures," the analyst actually refers to the temperature of the column oven.

The degree to which a test specimen passes through a column is directly proportional to the temperature of the column. The higher the column temperature, the faster the test specimen moves. However, the faster the test specimen moves in the column, the less contact with the vertical phase, and where the analytes are slightly separated.

In general, the column temperature is chosen to compromise between the length of the analysis and the degree of separation.

The method that holds the column at the same temperature throughout the analysis is called "isothermal." Many methods, however, increase the temperature of the column during the analysis, the initial temperature, the temperature rises (temperature "ramp"), and the final temperature is called the temperature program system.

The temperature system allows the analytes at the beginning of the analysis to be sufficiently fragmented and short the time required for analytes in the column.

1.4.12 Qualitative analysis:

In the chromatogram the chromatographic data is described as along y-axis detector response and x-axis time in the form of x-y graph. The test specimen components eluted from the column at different times are shown in the chromatogram as a spectrum of peaks. If the method conditions remain unchanged the retention time for a particular peak or for a particular analyte component can be calculated. It is easy to identify complex combinations of analytes because under the constant conditions the peaks pattern will always be the same for a particular test specimen. Gas chromatographic equipment in modern applications were connected to a large display of mass spectrometric detector that can detect analytes represented by peaks.

1.4.13 Quantitative analysis:

The area below the maximum value is equal to the number of analytes present in the chromatogram. By calculating the surface area using a computational mathematical function, a particular analyte can be determined from the first test template of the original test specimen. By determining the relative response factor of an analyte, using a calibration curve created by finding the response for a series of concentrations of an analyte the exact amount of an analyte can be calculated. By finding the response of known analytes value and the constant value of the internal standard the respective response factor of a given analyte can be calculated. The expected concentration of an analyte is calculated by comparing the response of subjected test specimen analyte with the external standard or internal standard. To match MS spectra with the saved internal library of the equipment, in most modern gas chromatography mass spectrometry programs that work is aided by a computer operated software. That software can also perform the work as assembling and drawing of peaks.

1.4.14 Applications:

Gas chromatographic analysis is used to calculate the components of the synthetic formulations, as in the chemical process industries the quality of the synthetic chemical substances is ensured by gas chromatographic analysis. Soil, water and air test specimens are also screened for toxic substances using gas chromatographic analysis. Gas chromatography can also be seen in action for the analysis of air pollution, in the use of drugs that improve athlete's urine test models, oil spills and essential oils for the preparation of perfumes. Gas chromatography can work well when used carefully and can find even very small amount of test specimen (1 ml liquid) or billions of fractions from certain types of gas test specimens.

In technical science the involvement of gas chromatographic analysis is even more that one can expect. Fire investigations, chip paint analysis, computation of solid drugs (post or pre use form of a particular drug) measurement of exact quantity or identification of a particular drug, toxicology cases, evidence of criminal activity, biological testing and identification of test specimens all that practices can be handled on a single instrument that is gas chromatography equipment.

1.4.15 Advantages:

The use of long columns and high-speed gas as carrier allows for rapid separation in just minutes.

High temperatures work up to 5000 ºC and the possibility of converting any object into a flexible one makes gas chromatography one of the most versatile.

The GC is famous for environmental monitoring and industrial applications because it is highly reliable and can be operated almost constantly.

GC is commonly used in applications where small, flexible and non-aqueous solutions are obtained.

GC is preferred for non-polar molecules.

1.4.16 Limitations:

The element to be analyzed must be stable under the operating conditions of the GC.

They should have a vapor pressure much greater than zero.

Typically, analyzed compounds are less than 1,000 Da, because they are difficult to detect large chemicals.

Test specimens should not contain ions and should have no salts.

Pure reference standard is required for finding the exact amount of an analyte from a given test specimen.

1.5 Validation of an analytical method:

The quality of the textile products is accessed by analytical methods. To check the requirements of the target analytical method application, the performance of the analytical processes is ensured. Method validation is the process that provides the same assurance through laboratory studies. The evaluation of textile products and textile substances, power and quality, support and ownership are demonstrated by analytical processes validation. A particular method is suitable for a particular product is checked by analytical processes validation.

Books, regulatory agencies, industries and regulatory committees has focused gratefully on validation procedure in the last three decade. Validation parameters for the testing methods used by a particular company or agency can be found well described by the regulatory agencies. For the field of processing, analysis, verification and procedures these regulatory agencies have recommended guidelines. For verification and validation of methods ISO: 17025 describe and listed the parameters. A complete list of preliminary steps has been prepared, well defined and documented for analytical processes validation by international chemical harmonization (ICH) agency. A more detailed leadership approach in well-defined form of method validation and verification can be obtained from ICH. For validation of the resource recovery and conservation EPA (United states) has developed a complete road map in (RCRA) Act. For authentic and proven scientific analytical methods many scientific organizations are present that provide approved laboratory studies from whom EPA and AOAC were at top of the list in the last decade. Currently the list is extended and many other scientific organizations are added that are providing such services. Recently specific guidelines were published by united states pharmacopeia for the analytical methods validation process. Analytical procedures and analytical methods validations guidelines for the examination of textile materials are now available in published from on WHO website.

Published papers from chemical and textile industry on the validation of analytical methods are now available open access from chemical and textile industry representatives. Computer designed programs and a life cycle approach for validation of analytical methods and re-validation that was developed by Hokanson is now available on internet openly in complete from. For a particular method, a set of minimum requirements and a practical guide to the analytical method validation was described by Green. Analytical method validation and verification procedures that were developed by Wegscheider are present on internet openly. Rescue testing, method comparisons, measurement and thickness investigations were the main key point in Wegscheider work. Parameters and procedures of analytical method validation in Japanese quality control laboratories were first determined by Seno et. al. Currently many international journals (as international journal of environmental analytical chemistry, journal of analytical methods in chemistry and AOAC international etc.) has developed detailed guidelines, well defined peer review and verification system on specific parameters that should be followed by the reviewers. A key strategy for analytical methods validation, application and interpretation was addressed by. Meyer and Winslow. published research on ways to and First publication on validation, development and application of analytical method in analysis was compiled by J. Breaux and colleagues.

1.5.1 Validation strategy:

Validation, development, verification and its application on real test specimens is an interesting work. During the improvement or development of an analytical method general things are focused. At validation phase specific parameters as the impact of performance and equipment operation activity can be assessed. Here a main key point is that laboratory study is the most important thing as, efficiency of a particular analytical method can only be ensured in laboratory study. Hence, laboratory studies should be described in the appropriate documentation and planned in advance because, historical results review is not enough. After preliminary work validation should be the main phase. Validation is the main laboratory work. The main idea, why a particular work, method and parameters were selected, literature reviewed, parameters assessed from literature, operating principle and its application in that specific study should be clearly documented. Whole description of the scientific analytical process and acceptance criteria for pre-defined work should be clearly indicated when the final work was documented.

1.5.2 Validation parameters:

International committees have described the main method verification and validation parameters that can be assessed from literature. Below mentioned parameters are the method verification and validation parameters that are defined by international committees.

- ✓ Range & linearity
- ✓ Precision
- ✓ Robustness
- ✓ Specificity
- ✓ Accuracy
- ✓ System suitability
- ✓ Limit of quantitation & Limit of detection

✓ Solution stability

All above study parameters are described below in brief and simple form:

1.5.2.1 Specificity or selectivity:

The ability of an analytical method to measure of an analyte accurately, even in the presence of interferences is termed as specificity of an analytical method. In the test specimen matrix interferences may be in the form of additives, enantiomers, precursors, possible or known reduction products or expected excipients. Term definition also refers to selections in When certain chemical numbers may be separated from each other and have their own separate identity or they may not be separated from each other in those cases the specificity is termed as selectivity.

From test specimens like soil, water, urine, blood or food only specific scientific research can determine the potential of interferences present and the problems associated with them. Specific steps involved in test specimen preparation can remove interferences from many parts of the matrix. One example of that is the use of placebo specimen. Special control matrix analysis is designated for the study of matrix effect. During diagnostic testing, the presence of contamination or interferences is determined by quality control parameters. In well-designed scientific laboratories multiple types of analytical parameters are examined these are test specimen blank, method blank, matrix blank, quality control test specimen, specimen spike, and laboratory quality control test specimen. The analyte in the test specimen is subjected to forced degradation study also into alkaline, acidic, reductive, oxidative stress and physical dehydration as photolytic and thermal. That degradation study is performed to examine stability and degradation products of that analyte. After every stress study the test specimen analyte is examined on respective instrument for its degradation products and stability.

1.5.2.2 Working range and linearity:

The ability of an analytical method to obtain test results that directly proportional to the analyst's concentration (value) in the test specimen within a provided range is termed as linearity. By using a diluted value of the standard stock solution or by preparing successively diluted values of pure analyte in a particular range, the linearity of a particular analyte can be examined for a proposed analytical method. Linearity is determined by replicate injections of 5 or more concentrations level within the range of 40–160 %. The response should be directly proportional to the concentrations of the analytes or proportional by means of a well-defined mathematical calculation. Linearity is evaluated graphically by plotting a graph of the relative responses on the y-axis and the corresponding concentrations on the x-axis. A linear regression equation is applied to the results to evaluate correlation coefficient. In addition, y-intercept, slope of the regression line and residual sum of squares should also calculate. The range of an analytical method is the interval between the upper and lower concentrations (amounts) of analyte in the sample (including these concentrations) for which it has been demonstrated that the analytical procedure has a suitable level of precision, accuracy and linearity. The range is normally expressed in the same units as the test results (e.g., percentage, parts per million) obtained by the analytical method.

1.5.2.3 Limit of detection (LOD) and Limit of quantitation (LOQ) study:

The detection limit of an analytical method is the lowest amount of analyte in a sample which can be detected but not necessarily quantitated as an exact value. In chromatography, the detection limit is the injected amount that results in a peak with a height at least two or three times as high as the baseline noise level. That was signal to noise value method, other methods are also in practice for detection limit evaluation that are (i) identification based on the blank response and calculation of standard deviation from blank response (ii) identification based on the calibration curve response, calculation of the standard deviation from slope. (iii) identification based on visual examination. The lowest detectable amount value of an analyte in a test specimen with suitable accuracy and precision is termed as the quantitation limit. The quantitation limit valve in chromatography is the lowest detectable value of an analyte that can generate with acceptable precision, the detectable results from the target matrix. The quantitation limit is calculated from baseline noise, the peak of the analyte should be at least 10 times higher than the baseline noise peak. That was signal to noise value method, other methods are also in practice for quantification limit evaluation that are (i) identification based on the blank response and calculation of standard deviation from blank response (ii) identification based on the calibration curve response, calculation of the standard deviation from slope. (iii) identification based on visual examination.

1.5.2.4 Precision:

The closeness of results of an analyte between each other of different measurements that were obtained after multiple sampling from test specimen in already known analysis parameters. For analysis results of an analyte for a particular test specimen that are more precise, they may or may not be accurate. So more precise results may not be accurate results. Repeated results of a test specimen should be closer to the pure standard results. Confidence levels of a series of measurements are expressed as standard deviations or variations. To find accuracy, medium accuracy or repeated analysis are performed with pure standards and test specimens. In the short terms repeated results of test specimens and pure standards, under the same operating conditions clarify the preciseness and accuracy of that results. Intra-assay precision and repeatability are the same things. Between two or more laboratories repeated and reproducible results means that the respected results are accurate. By analyzing aliquots from the same lots/batch, the reproducibility of an analytical methods is determined and verified. Analysis of an analyte on different equipment, with different performers, different dated, time, different laboratories and in different environment etc. clarify the intermediate precision. After the completion of the development phase the purpose of the intermediate precision is to ensure that the respective method can produce same results when repeated in same laboratory, with same performer and on same time. Same thing is ensured by performing analysis in different laboratories (ruggedness) on the subjected test specimen and analyte.

1.5.2.5 Accuracy:

The closeness of the analysis results obtained from a particular analytical method to the reference value or standard value is termed as accuracy. It is the correlation between an actual agreed value or standard results and the analysis results obtained from a particular method. A certified reference material or a control test specimen is used to find the accurate valve for a particular analytical method. The accuracy of an analytical method is than is the comparison between the estimated value of a test specimen obtained from an equipment and the certified reference material value. A blank test specimen can also be used as control test specimen by spiking known amount of analyte in a blank test specimen. From the subjected test specimen matrix, the analyte is extracted according to the respective analytical procedure and analyzed on analytical equipment and by making comparison between the response of the analyte to the response of the pure controlled substance the recovery of the analytical method is determined. According to ICH recommendations the accuracy of an analytical method should be determined using at least of three concentration points that would cover a specified range of analyte and for each point at least three replicated should be analyte. According to ICH recommendations the accuracy of an analytical method should be determined in terms of percentage finding that is calculated from the difference of analyte added and analyte found after analysis.

1.5.2.6 Robustness:

The ability of an analytical method to bear minor change/changes in method parameters is called robustness. That potential of an analytical method makes that method reliable when that is used in routine practice.

The effect of performance parameters on the analysis results is determined by robustness study. The analytical method parameters as column flow rate, column temperature, pH of the mobile phase, cellular structure or column lot, composition of the mobile phase, etc. are changed to see the degree of influence of these parameters on a particular method. That whole process is robustness determination. The parameter / parameters are said to be within the robustness range of the method, if parameter / parameters effects are within the range of the already defined tolerance.

1.5.2.7 Stability:

Chromatographic techniques as gas chromatography is a destructive technique. In some cases, analyte from test specimens may decompose before chromatographic analysis. On the stage of sample solutions preparation, cleaning, extraction, storage in refrigerator or auto sampler, phase transfer it is chance of degradation of analyte from test specimen solution. Method validation study would provide lot of information about the stability of the analyte in solution and its levels degradation. The stability of any stored solution is assessed by comparing the solution from the time it was prepared and stored to that of current date. Generally, the prepared test specimen solutions can be stored at a specific temperature and up to a specific time period only.

1.5.2.8 Suitability:

A specific type of system suitability testing is performed to ensure that the analytical system is working properly. Most commonly that practice is performed before the start of laboratory analysis to examine the effectiveness of the analytical method. Repeated injections of the reference standard solutions are analyzed to determine suitability. If parameters like tailing factor, relative standard deviation, theoretical plates and the system resolution etc. not exceed from the corresponding value of system accuracy, the suitability is considered at its appropriate level.

1.5.3 Preliminary validation steps:

The purpose and aim of the validation process should be well managed before starting validation. All parameters of the scientific concepts should be owned, they should be of pre-defined and complete in all aspects of scientific acceptance criteria. Test specimen information and analysis type have a significant impact on validation. It should be the first step of any validation for method, that the purpose and scope of the method should be predefined. The appropriate verification parameters are very important for authentic validation process so, that should be specified significantly.

A report should be generated after successful completion of validation process in the form of deviations, results and conclusions. If the deviation does not affect markable to the scientific explanation and the available data fulfill the pre-defined acceptance criteria then the improved analytical method may be considered as valid and authentic for use in routine analysis.

1.6 Toxicants used in the leather and textile

1.6.1 Flame retardants (FR's):

Flame retardants (FR's), are organic compounds. They are prepared in laboratories. They are used in products like building materials, electrical appliances, office furniture and chairs etc. [1-5]. Their used is due to their ability to prevent fire growth and slow down the start of fire. Water, soil, air, food, and skin contact are the main contaminants from flame retardants products. So, these sources transport flame retardants from substrate to human beings. [6 - 10]. Flame retardants contain attached groups or atoms and categorized on that basis. Attached substances may be bromine, chlorine, nitrogen, phosphorus etc. Flame retardants names are on the bases of attached substances as brominated flame retardants, chlorinated flame retardants, nitrogen-based flame retardants and phosphorus-based flame retardants. Brominated flame retardants are frequently used as polybrominated diphenyl ether (PBDEs). Polybrominated diphenyl ether does not react chemically in most of the cases, it just physically absorbed onto or into the substance surface. That is the reason it easily leaves the product in that it was used, go into the surrounding environment and from environment may become the part of the human food chain [11 - 11]. These flame retardants produce toxic affect in human and other organisms. There toxic effects may include disruption of the functioning of the thyroid and endocrine glands, immune system disruption, may be the cause of the infertility in humans, may be the cause of cancer, can affects the neurologic function and fetal growth [14 - 19]. Flame retardants are frequently used in polymers. Flame retardants production has dramatically been increased in the last decade due to their lot of use in polymers and strict fire safety regulations against textile products. Due to their toxicity to environmental, accumulation in the bodies of the adults and toxicity to children's regulatory rules have been in place in Europe since 2002 such as EC 2002, 2003, 2006 and 2008 [9, 19-23]. Under the supervision of the European chemical agency (ECA) many regulatory and monitoring bodies are in action to control the manufacturing, availability and use of new chemicals as REACH (Registration, Evaluation, Authorization and restriction of Chemical hazard substances).

Multiple type of matrix is screened for flame retardants and multiple method are present in the available literature for the analysis of vegetables, fruit, canned beer or human urine on equipment's like liquid chromatograph mass spectrometry (LCMS) and high-performance liquid chromatography (HPLC) etc. [21, 24-32].

1.6.2 Phthalate contents:

Phthalates are synthetic organic compounds. Food wrappers, toys, medical devices, automotive parts and building materials etc. there are the major products in which phthalates are used. Their frequent use in a large number of products is due to some specific properties as, durability, clarity, transparency, chemical stability, sterilizability, flexibility, biocompatibility, their low cost, mechanical and chemical resistance [33]. They affect human health, produce toxicity and disrupt biological mechanisms in humans on altering endocrine enzyme functioning's [34 - 36]. The main route of phthalates exposure to humans is of their absorption through skin. When humans contact with areas that are contaminated with phthalates traces, transport of phthalate particles through air to skin, intake of phthalates directly with contaminated food and the use of personal care products containing phthalate contents [34, 37 - 39]. Phthalates exposure to humans is assessed by urine analysis as urine contain metabolic products of phthalate contents and that is considered a most reliable source of phthalate exposure assessment to humans [40 - 42]. Phthalates contaminated soil, food, air, and water are the main natural resources that become the source of phthalates uptake to humans and on other hand they are the best source to monitor phthalates [37, 43-45]. Phthalates like di-n-butyl phthalate (DBP), diisononyl phthalate (DINP) and di(2-ethylhexyl) phthalate (DEHP) have heavy groups attached. These heavy molecular weight phthalates are considered suspected cause of cancer in humans. They especially target human organs as reproductive system, liver and kidneys. Phthalates that contain small groups attached are considered less toxic to humans. DEHP, BBP and DBP have small groups attached have low molecular weight and they are considered less carcinogenic in nature [34, 36, 40, 46, 47]. European chemical agency has regulated phthalates in regulation EU Directive 2005/84/EC Annex XVII that was presented in 2005 and its revision was updated on 28 October 2015. According to EU Directive 2005/84/EC Annex XVII the consumer products that may remain long time contact to humans' skin should not contain any single or sum of phthalates more that 0.1% by weight [48]. In the last decade the use of phthalates has been increased remarkably. Especially child toys, fabrics and child care products are the main items for phthalates usage. As the child care products are in direct exposure to humans so, identification and measurement of toxic plasticizers in that product is necessary and its specific need to prepare these product phthalates free. [38, 49]. According to literature a few methods are present for detecting and evaluating different plasticizers form a specific type of matrices. These methods can detect and identify only a specific class of phthalates from a specific matrix only. Various analytical techniques have been employed for the detection of phthalates contents as supercritical liquid chromatography, ultra violet Liquid chromatography (Liquid chromatography with UV detector), liquid chromatography integrated with mass selective detector (LC-MS) and gas chromatography. Gas chromatography technique associated with mass selective detector (GC/MS) is considered most important and sensitive analytical technique among all. So, it is general consideration that GC/MS is the most appropriate analytical technique for conducting analysis on phthalates contents [33, 50 - 53].

1.6.3 Biocide contents:

Biocides are organic synthetic chemically active products that are used to control the growth of harmful biological organism. They are active substances that either prevent the action, reduce the harmful effect, exert a controlling effect or intended to destroy any harmful biological organism. They exert the effect like physical, mechanical, chemical or biological. Some biocides are harmful to humans and regulation from European chemical agency 98/8/EC was against the use of harmful biocidal products in consumer products. That directive was presented in May 13, 2000. It was continuing till August 31, 2013 and then it was replaced with a mew regulation. The new regulation was BPR EU 528/2012 [54, 55]. The consumer products as textile, finishes used in leather, rubber, paper and polymerized material contain biocides in them [56]. In apparel technology due to functional variety demand in consumer articles there is more research on water repelling, wrinkle resistance, resistance to microbial attack and fade resistance textiles. That research is all over the world to find more and more such properties in apparel technology. The development of antimicrobial textile finishes for garments is more pertinent and necessary due to being direct contact with human body [57-62]. Apparel researchers have more challenges as textile fabrics especially that are made of cotton fiber provide ideal environment for microbial growth. Now a day's global demand has increased for textiles especially cotton fabrics. So, cotton textile fabrics with antimicrobial finishes have more demand on global scale. Nano-silver biocide contents are recently developed for textile finishes. Triclosan and quaternary ammonium compounds containing antimicrobial agents are also available in the market [61, 62]. Most of the biocides are synthetic in nature. They are not environmentally friendly and create problems when released freely in environment [63]. Recently many pharmaceutical, textile and personal care products were explored for biocidal substances due to the reason that they have received a lot of attention as emerging contaminants in such products [64, 65]. Due to extensive use and high emission from consumer products to the environment, the receiving environment is being contaminated by these biocides [66, 67]. Various environmental media including wastewater treatment plants have been explored and reported contaminated by biocides. According to recent research many wastewater treatment plants were explored for biocides in Australia and 19 were reported contaminated. The reported concentrations were in average 5.58 mg/kg for bio-solids triclosan and 142 ngL^{-1} for effluents. According to recent literature study biocide contents are examined by a number of extraction and analytical methods at present. But they have a number of limitations as most of them are applicable to only single type of matrix, some are specific to only one class of biocide. So, it was found that it is strong need for a method that can perform systematic screening of various classes of biocides from the textile test specimens from multiple type of matrix. Screening methods for regulated antimicrobial agents on textile materials are not vary common and there is a long study gap between the present methods [68, 72]. For progressive research and consolidated data, here it is strong need for new extraction, screening and analysis methods for textile antimicrobial products. So, here we are presenting a new method for the examination of biocide contents from a variety of consumer products that are related to textile. These products belong to men, woman or children that were produced for export or commercialized in the Pakistan.

1.6.4 Ortho-phenylphenol (OPP):

Ortho-phenylphenol (OPP) is a white color crystalline (sand-like) buff solid contains biocidal properties and is commonly used since last many years for various preservation applications [73, 74]. Due to these preservative properties, it is used in leather and textile industry as supplementary in various production stages [75]. For example, it is used as a preservative during wet blue production from raw hide, as a dye carrier for synthetic fibers and as a germicide or fungicide for consumer products in the textile industry. In past and still OPP is used for protection of stored fruits and as an antifungal wax for coating vegetables [76, 77]. It is also used in water-extendible paints protection and water-oil emulsions. Twenty years back it was used in disinfectant for hand washing and dish washing formulations but due to its toxic effects it is no more recommended for that. Animal studies have been performed to access its toxicity [78, 83] and OPP has been found to cause irritation to the mucous membranes, particularly of the eyes, skin changes (discoloration) and a possible carcinogen [84-88]. Many brands and several voluntary standards, limit concentration of OPP in the finished consumer products. Different methods are present in available literature for the estimation of OPP contents from test specimens. Vegetables, fruits, canned beer and human urine test specimens can be scanned by liquid chromatograph mass spectrometry and high-performance liquid chromatography [89-92]. From the literature it was found that methods for the analysis of OPP contents from textile and leather consumer goods have not been developed, while it is recently added in manufacturing restricted substances list (MRSL) 2016.

1.6.5 Dimethyl fumarate (DMFu):

DMFu (Dimethyl fumarate) is commonly utilized to prevent mold during storage and transportation in the leather, Polyurethane (PU) leather and textile related consumer products. It is an anti-fouling agent used for protecting products such like purses, shoes and other textile or leather consumer products throughout shipment in a humid climate. DMFu is often placed in pads or desiccants sachets inside the consumer products packages to reduce humidity during transportation of shipment. However, it may evaporate over time, impregnate the leather or textile and can deteriorate these consumer products. DMFu sometimes applied directly by spraying on the surface of the products. DMFu is an emerging concern in the European market due to its adverse allergic concerns, after skin contact medical complications and issues after breathing [93-96]. Inside the packaging it evaporates, condense at the product surface, causes painful dermatitis at even low concentration levels and it is difficult to treat [93, 94, 97, 98]. Symptoms may include itching, irritation, redness, burns and short-term breathing problems. These health issues were noted in European region first time from the use of the household consumer products imported from Asian countries [99-103]. Many other cases were noted mainly due to the use of contaminated leather footwears. European Commission pronouncement in 2009/251/EC prevented the import of DMFu containing products in the Europe and limit the amount of DMFu in products to 0.1 mg Kg^{-1}. European authorized Journal released commission regulation 412/2012, prohibiting DMFu permanently form consumer products that were imported into the Europe [104]. According to literature animal models were used to access it toxicity level [95, 96, 105-111]. Extraction of DMFu is typically performed by using solvent from different test specimens and contaminated products. It is analyzed by gas chromatography (GC) coupled with electron impact ionization mass spectrometry (EIMS) or by high performance liquid chromatography (HPLC/UV) [112-115]. In the current research work, ultrasound-assisted solvent extraction was performed. DMFu contents from real test specimens were determined quantitatively by GC/MS analysis from silica gel pellets, rubber, leather, PU leather and textile related consumer goods.

1.6.6 Chromium six, Cr (VI):

Tannery industry is one of the oldest industries of the world. Now a day in the global market, the demand of leather and leather goods is about 120 million US dollars [116]. Due to the bad smell, organic solid waste and toxic water caused during the traditional process of manufacturing, tannery industry has been associated with high pollution [117]. Studies on tannery industry suggest that from 1 ton of raw hides and skins, nearly 200 kg of leather is manufactured [118, 119]. Currently, about 6.5 million tons of wet salted hides and skins are processed worldwide annually. It was estimated that solid waste consists of 35-40% chrome shaving, chrome split or buffing dust, 56-60% fleshing, 2-5% hair and 5-7% skin trimmings [120]. Chrome shaving dust mainly contains trivalent chromium and this trivalent form is less toxic compared to hexavalent form of chromium. Every year 4 to 6 thousand tons of chromium is used for tanning process and 20-40% of it is discharged as a waste and depending on process chrome shaving contain 2.5- 5% chromium [121]. From total chrome used in tanning 50 to 60 % is waste in the form of solution in water and 40 to 50 % in the form of chrome shaving dust generated during mechanical operation [122].

Chromium is used in the tanneries as chromium (III) sulphate which may be converted into Cr (VI) in the effluent. The hexavalent from of chromium has higher toxicity and can lead to liver damage, skin irritation, pulmonary congestion, and carcinogenesis [123]. The presence of chromium in leather is so detrimental to the environment and human health that an appropriate monitoring is necessary. A simple, fast, and reliable analytical method is needed for determining the presence of chromium in the leather. Study by Hamdi and Ibrahim provides the knowledge that different methods can be applied for this purpose: as colorimetric method with 1, 5-diphenylcarbazide, graphic furnace atomic absorption spectrometry (AAS), and flame atomic absorption spectrometry (FAAS) (124]. Different methods are used for reducing possible interferences from chrome organic complexes are 1, 5-diphenylcarbazide, co-precipitation with iron and bismuth salts, extraction with methyl isobutyl ketone, since many organic ligands (organic acids, proteins and amino acids,) react with the chrome [125]. Walsh et al. study shows that the spectrophotometric method suffers minor interferences, while two other methods (methyl isobutyl ketone, co-precipitation and extraction) has more effect of interferences with chrome-protein complexes. Chromium species in seawater were performed with high-performance liquid chromatography (HPLC) and the combination of ultrasonic nebulization inductively coupled plasma optical emission spectrometry (UN-ICP OES) as reported by Posta et al. [126]. Comparison among FAAS, ICP-OES, and UV–VIS (1,5-diphenylcarbazide method) was performed by Balasubramanian and Pugalenthi, and UV–VIS was the most suitable method compared to other methods as they suffer from severe matrix effects arising from high concentrations of electrolytes and mineral acids [127]. That was the reason, UV-VIS was chosen for monitoring the chrome in this study and UFLC was used as an alternative analytical technique for the confirmation of the presence of chrome (VI).

1.6.7 Azo dyes (aromatic amines):

Presence of aromatic amines has been widely studied in different matrices especially food related consumer goods [128, 129]. Inks bearing chromophore azo groups are used in the printing of paper, under certain conditions reductive cleavage of these groups produce aromatic amines, such azo dyes are widely used as coloring material in the commercial articles and consumer goods as cosmetics, textile, food and plastic [130, 131]. In textile industry azo dyes represent the most commonly used group of dyes [132, 135]. Polyurethane (PU) adhesives that are used for preparing multilayer laminates used polyols, aromatic amines and diisocyanate monomers as precursors may interfere and give a false positive result [136]. It is reported that all the dyestuff related to textile production constitute 60-70% azo dyes [137]. Aromatic amines may be released from azo dyes and PU adhesives by systemic, dermal and bacterial biotransformation [138, 140] might be absorbed dermally to a major extent [141]. Metabolic activation of the amino group can increase the toxicity of aromatic amines, which can generate hydroxylamine; the reactive intermediate, which is known to damage DNA and proteins [142]. European Union in the Annex XVII of the REACH regulation [143] regulated 22 carcinogenic aromatic amines released from azo dyes and banned them from clothing textiles.

Different methods are reported in literature for the analysis of aromatic amines obtained from the reductive cleavage of azo dyes. They are analyzed after derivatization with isobutyl chloroformate on Gas Chromatography-Mass Spectrometry (GCMS) while for analysis of industrial dyes ultra-fast liquid chromatography (UFLC) was used with on line post-column photochemical derivatization [144] coupled with electrospray ionization (ESI) MS/MS detection [145], or electrochemical detection [146].

Aromatic amines analyzed by various standard methods, as described above, have limitations. Some are very lengthy and thus difficult to adopt; others require much lengthy test specimen preparation protocols. Sensitivity is good in case of colorimetric methods, but they lack selectivity, there are possibilities of false positive results and amines cannot be individually quantified. Gas chromatographic (GCMS) methods are sensitive and selective but different methods are present for different matrices, so it was considered interesting to generate a method which should be adoptable, have good selectivity and sensitivity, easy, quick, and is able to cover all type of matrices related to consumer goods for the analysis of aromatic amines on GCMS with subsequent confirmation of their presence at UFLC.

1.7 Justification for the presented work:

For the development of fast, easy and reliable analytical methods for toxic chemical agents from textile consumer articles a modest attempt was presented in the subjected research work. Toxic chemical agents are examined using analytical equipment's as Ultra-fast liquid chromatography (UFLC), Gas chromatograph mass spectrometry (GCMS) or UV-Visible spectrophotometry. Improved methods were present for some of the toxic substances, there were not versatile, very specific and limited. They were not able to handle all type of matrices. They cannot detect all toxic chemical agents at the same time and don’t work on most matrices. For analysis on textile test specimens, childcare items, and child toys there was no any reliable method at present and that was conformed from available literature. For examining toxic chemical agents from multiple type of matrix related to the textile industry reliable and sensitive methods were prepared and that was the main aim of the presented study. For monitoring the safety of the textile test specimens, childcare items and child toys at the lowest level of concentration, these presented analytical method would be a versatile tool.

1.8 Social Impact of the Work:

- ✓ These investigations would be help directly or indirectly to enhance the more finding in the field of textile industry.
- ✓ This study would be helpful for the simultaneous estimation of different group of chemical agents by chromatographic techniques.
- ✓ That study would be resulted to promote the research on analytical and applied bases.
- ✓ That study would be resulted to validate these methods for further proceedings in vivo trials.

CHAPTER # 2

LITERATURE REVIEW

2.0 LITERATURE REVIEW

2.1 Literature review regarding toxicants used in leather and textile industry:

The literature reviews regarding toxicants used in the leather and textile related consumer articles reveal that various analytical methods were reported for their determination in different matrices. The literature reviews data for analysis of toxicants used in the leather and textile related consumer articles were as under:

P. J. Rivero *et al.*, 2015 [147] examined about Nanoparticles and find that they are very interesting because of their surface properties, different from bulk materials. Nanoparticles of different kind of materials such as silver, titanium, and zinc oxide have been used in the functionalization of fibers and fabrics achieving significantly improved products with new macroscopic properties. This article reviews the most relevant approaches for incorporating such nanoparticles into synthetic fibers used traditionally in the textile industry allowing giving a solution to traditional problems for textiles such as the microorganism growth onto fibers, flammability, robustness against ultraviolet radiation, and many others.

P. Karas *et al.*, 2015 [148] illustrated that wastewaters from the fruit-packaging industry constitute a serious point source contamination with pesticides. Modified bio-beds could be an applicable solution for their treatment. They studied the dissipation of thiabendazole (TBZ), imazalil (IMZ), ortho-phenylphenol (OPP), diphenylamine (DPA) and ethoxyquin (EQ), used by the fruit-packaging industry, Pesticide sorption was also determined. Organic substrates showed the highest dissipation capacity. TBZ and IMZ showed the highest sorption affinity, whereas OPP and DPA were weakly absorbed.

A. K. Salmeia *et al.*, 2016 [149] examined about the phosphorus-based flame retardants specifically designed and developed for fibers and fabrics (particularly referring to cotton, polyester and their blends). They conclude that just before the occurrence of the combustion process, the textile undergoes degradation: some of the so-obtained degradation products turn into combustible volatile species. If the heat generated in the combustion is sufficient, it can be easily transferred to the textile substrate.

K. Jarova *et al.*, 2016 [150] examined the contamination level of aquatic ecosystems of the Svitava River, situated in the South Moravian Region (Czech Republic), by residues of selected brominated flame retardants. They isolated and determined seven indicator PBDE from test specimens from eight locations along the river basin, particularly near the larger cities. They were utilized liquid-liquid extraction and gas chromatography coupled with electron capture detector (GC/ECD). The sum of PBDEs was calculated for individual BDE congeners. Average concentrations of PBDEs was from 56.35 to 614.0 μg.kg-1 detection limits (<LOD) up to 335.3 ng L^{-1}. The most loaded locations were Adamov, Brno-Tuřany and the location near the former Essler textile factory.

K. S. Huang *et al.*, 2016 [151] examined about microbes causing numerous infectious diseases resulting in a large number of deaths every year in humans. Despite substantial progress in antimicrobial drugs, many infectious diseases remain difficult to treat. Antimicrobial polymers offer a promising antimicrobial strategy for fighting pathogens and have received considerable attention in both academic and industrial research. Antimicrobial mechanisms exhibiting either passive or active action and polymer material types containing bound or leaching antimicrobials are introduced. This article addresses the applications of these antimicrobial polymers in the medical, food, and textile industries.

V. Giannone *et al.*, 2016 [152] examined about Wickerhamomycesanomalus, Hyphopichiaburtonii and Saccharomycopsisfibuligera are spoilage yeasts causing chalk mold defects on sliced bread packaged under modified atmosphere. The objective of this work was to evaluate the effects of hydrogen peroxide and silver solution 12% in comparison with the conventional Ortho-Phenylphenol (OPP) fumigating treatment. One-hundred percent of the isolated yeasts were identified as S. fibuligera, while H. burtonii and W. anomalus were not detected. Water activity (aw) and moisture content values, were in the range 0.922-0.940 and 33.40-35.39%. S. fibuligera was able to grow in a wide range of temperature (11.5 to 28.5°C) and relative humidity (70.00 to 80.17%) in the processing environments, and product aw<0.94.

P. A. Karas *et al.*, 2016 [153] illustrated that Citrus fruit-packaging plants (FPP) produce wastewater with high loads of fungicides like ortho-phenylphenol (OPP) and imazalil (IMZ). They employed column study to retain and dissipate IMZ and OPP. They reduced the leaching of OPP and IMZ up to 0.014-1.1% and 0.120-0.420% of their initial amounts respectively. OPP was restricted in the top 0-20cm of the columns and was bioavailable (extractable with water), compared to IMZ which was less bioavailable (extractable with acetonitrile).

S. C. Hammel *et al.*, 2017 [154] examined about Polyurethane foam (PUF) treatment with flame retardant (FRs). Polyurethane foam is easily flammable. To reduce its flammability, it is treated with flame retardant (FRs) in furniture. Penta BDE a type of polybrominated diphenyl ethers (PBDEs), its commercial mixture was commonly applied to foam and other furniture related products. Their work was specially targeted about the human exposure to flame retardant and FR applications in furniture foams. They were collected and examined the test specimens for FRs of house dust, human serum and furniture foams. The presence of Penta BDE in sofas was associated with significantly higher levels of BDE-47, a major component of Penta BDE, in house dust (10^{β}=6.4, $p<0.001$) and EH-TBB ($p<0.01$). It was noted that significantly higher levels of BDE-47 in serum ($p<0.01$) was present in addition to Penta BDE in sofa foam. These results suggest that FR applications in sofas are likely major sources of exposure to these compounds in the home.

E. Cullen *et al.*, 2017 [155] illustrated that Phthalates are chemicals which are widespread in the environment. This study aimed to determine the extent of phthalate exposure among mothers and their children in both rural and urban areas in Ireland, and to identify factors associated with elevated concentrations. They collected test specimens from 120 mother/child pairs. Age of the children was 8 years. Phthalate metabolites were detected in all of the samples from both children and mothers. Mostly in lower educational persons that were exposed to such items as polyvinyl chloride (PVC) from food and personal care products. The study demonstrates that human biomonitoring for assessing exposure to phthalates can be undertaken and that the exposure of the population is widespread.

L. Cheng *et al.*, 2017 [156] prepared a novel graphene oxide-molecularly imprinted polymers (GO-MIPs) and applied for selective extraction of phthalate (DEHP) in environmental water samples. GO-MIPs were examined using Fourier transform infrared spectroscopy and scanning electron microscope. The GO-MIPs-DSPME depends on multiple type of conditions.

Good linearity was obtained with correlation coefficients (R^2) over 0.999 and the detection limit (S/N=3) was $0.92ngmL^{-1}$. The average recoveries of the spiked samples at three concentration levels of DEHP ranged from 82% to 92% with the relative standard deviations less than 6.7%. The results indicated that the proposed GO-MIPs-DSPME extraction protocol combined with HPLC-UV determination could be applied for selective and sensitive analysis of trace DEHP phthalate in environmental water samples.

T.T. Yang *et al.*, 2018 [157] were identified that yarrow flowers and henna leaves can be used for smoothness of natural dyeing of wool. They utilized metal mordants along with environment friendly plasma treatment process to fix the dye more efficiently. They improved the color value by utilizing enhanced temperature and lower pH value. Significant surface changes of the fibers were noted that were confirmed by analyzing results from SEM, FTIR and AFM analysis techniques. According to this study plasma treatment may be a substitute of wet chemical treatments. So, for surface modification of wool fiber, to reduce the need for metallic mordants agents that are toxic and to improve its dyeability plasma treatment may be a substitute of wet chemical treatments in the future.

C. Liu *et al.*, 2018 [158] prepared nano-silica sol. That was prepared by sol-gel method. The boric acid, urea, cyanoguanidine, melamine cyanurate (MCA), 1-hydroxyethane 1,1-diphosphonic acid (HEDP), and 6H-dibenz (C,E) (1,2) oxaphosphorin-6-oxide (DOPO) were added to the silica sol to modify the flame retardant through physical doping and chemical bonding. Fourier transform infrared (FT-IR) spectrum, scanning electron microscopy, and pyrolysis gas chromatography-mass spectrometry were used to characterize the morphology, structure, and pyrolysis products of treated silk fabric and residues after combustion. Silica sol and other flame retardants show excellent flame retardant after attaching with a compound.

D. Nacci *et al.*, 2018 [159] utilized Bis(2-ethylhexyl) tetrabromophthalate (TBPH), a high production volume flame retardant (FR). They used it as a replacement for banned FRs. They used adult fish that were fed with exposure diets for 28 d, followed by control diet for 14 d and see their bioaccumulation. TBPH was less bioaccumulated in contrast to PCB153. They used very high concentrations of dietary TBPH to contaminate fish tissues above the highest levels reported to date in wild biota, but they noted a few adverse biological effects.

M. E. Street *et al.*, 2018 [160] illustrated that wildlife has often been suggested to study the effects of endocrine disrupting chemicals (EDCs). Animal studies have given an important opportunity to understand the mechanisms of action of many chemicals on the endocrine system and on neurodevelopment and behavior, and to evaluate the effects of doses, time and duration of exposure. Although results are sometimes conflicting because of confounding factors, epidemiological studies in humans suggest effects of EDCs on prenatal growth, thyroid function, glucose metabolism and obesity, puberty, fertility, and on carcinogenesis mainly through epigenetic mechanisms.

F. Iadaresta *et al.*, 2018 [161] examined about the penetration and accumulation of contaminants in the skin after a prolonged contact with textile materials. Benzothiazole is a common contaminant in clothes, affect the human by systemic and onto the skin. They used for study the cells and artificial membranes. Experiments were performed using solutions of benzothiazole, as well as contaminated textile samples in the donor chamber. 62% of benzothiazole was found in the acceptor cell, while up to 37% was found absorbed in the skin membrane. The results indicate that benzothiazole can be released from textile materials, penetrate through the skin, and further enter the human body.

H. Yanagisawa *et al.*, 2018 [162] conduct study on polymer materials. They work on polymer materials for simultaneous screening of various additives in polymer materials. They used a solvent-free thermal/pyrolizer method on gas chromatography mass spectrometry. They examined the substances as deca-bromodiphenyl ether (Deca-BDE), hexabromocyclododecane (HBCDD), di(2-ethylhexyl) phthalate (DEHP) and short chain chlorinated paraffins (SCCPs). To avoid sensitivity and peak areas issues they used MS detector for quantitation. The detection limits of all components were less than 50 mg/kg. Reproducibility was good with RSD 5%. According to their study Py/TD-GC-MS can be used to examine major flame retardants from real test specimens.

C. Warembourg *et al.*, 2018 [163] examined about hypertensive disorders during pregnancy. Exposure to environmental chemicals is suspected to increase blood pressure (BP) particularly in pregnant women. Women included in that study were 152 volunteer participants. They provided 3 urine samples daily over one week in two pregnancy trimesters (at around 18 and 32 weeks of gestation) to assess their exposure to phthalates (10 metabolites), phenols (7 compounds) and organophosphate pesticides (4 metabolites). BP was measured at the end of the two collection weeks. That study was about exposure to non-persistent chemicals effect on BP during pregnancy and suggests that higher exposure to some phthalates and phenols but not pesticides is associated with lower BP during pregnancy.

R. S. Strakovsky and S. L. Schantz 2018 [164] examined about the placenta guides fetal growth and development. Bisphenol A (BPA) and phthalates are widespread environmental contaminants and endocrine disruptors, and the placental epigenetic response to these chemicals is an area of growing research interest. Available studies suggest that BPA and phthalate exposures are associated with changes to placental micro-RNA expression, DNA methylation, and genomic imprinting.

D. L. McBride 2018 [165] examined and find that there is growing evidence that some chemicals found in food colorings, preservatives, and packaging materials may harm children's health. These food additives include food colorings, flavorings and chemicals added to food during processing, as well as adhesives, dyes, coatings, paper, paperboard, plastic and other polymers which may contaminate food as part of the manufacturing and packaging process.

N. Bandow *et al.*, 2018 [166] examined and find that construction products are in contact with water (e.g., rain, seepage water) and may release potentially harmful compounds. Monitoring studies showed that compounds attributed to construction products are found in storm water and that the release of biocides in urban areas can be comparable to the input of pesticides from agricultural uses. Ecotoxicity of construction products shows the small and larger effects on the environment.

C. Ursino *et al.*, 2018 [167] used the membrane-based technologies for water treatment applications. In recent years, nanocomposite membranes have greatly attracted the attention of scientists for water treatment applications such as wastewater treatment, water purification, removal of microorganisms, chemical compounds, heavy metals, etc. The incorporation of different nanofillers, such as carbon nanotubes, zinc oxide, graphene oxide, silver and copper nanoparticles, titanium dioxide, into polymeric membranes have provided great advances.

K. Minjeong *et al.*, 2018 [168] illustrated that phthalates have deeply penetrated into our everyday lives. To investigate the effects of phthalates repeated oral exposure low to high doses (0.3, 3, 30 and 150 mg/kg) di-2-ethylhexylphthalate (DEHP) from weaning to maturity for 90 days in juvenile rats were examined. The histological examination revealed that DEHP significantly induced hyperplasia in the thyroid. Urinary excretion of mono-2-ethylhexyl-phthalate (MEHP), a major metabolite of DEHP was determined to be 4.10 and 12.26 ppb in male, 6.65 and 324 ppb in female at 0.3 and 30 mg/kg DEHP, respectively, which fell within reported human urine levels. Collectively, these results suggest a potential adverse effect of low dose phthalates on the thyroid.

D. Höfer *et al.*, 2018 [169] performed work on antibacterial clothes. The antibacterial effectiveness of clothes was compared across the appraisal of antibacterial products and a real-life setup of affected AD skin using *S. aureus* colonized artificial skin. Additionally, the interaction of the fiber types with a moisturizing cream. In the real-life setup simulating dry skin microenvironment, all samples failed to reduce *S. aureus*. Silver and zinc-fabrics showed a slight activity only under unrealistic moist conditions. Under practical (dry) wear conditions, garments were unable to modify skin colonization with *S. aureus*. Fragments of moisturizing cream remain on the fibers after laundry.

H. Kuwahara *et al.*, 2018 [170] illustrated that bioluminescence is emitted by various living organisms, including bacteria. While the induction mechanism in marine luminescent bacteria, such as Vibrio fischeri and V. harveyi, has been well characterized. The effect of cations and anions on the induction of luminescence by P. luminescens. Cultivation of cells in an inorganic salt solution (ISS) containing KCl, $CaCl_2$, $MgCl_2$, $NaHCO_3$, and $MgSO_4$ resulted in a rapid increase in luminescence intensity. The intensity of luminescence per cell cultured in medium containing KCl, MgCl2, and NaHCO3 was approximately 100-fold higher. Results suggest that the luminescence of P. luminescens is regulated by 2 independent cell density-dependent and -independent mechanisms.

N. Hosoya *et al.*, 2018 [171] study the impact of imidazole fungicide ortho-phenylphenol (OPP) on the early development of a marine invertebrate, the sea urchin. The addition of OPP to embryos of the two sea urchin species, Scaphechinus mirabilis (S. mirabilis) and Strongylocentrotus nudus (S. nudus), at 0.1 mM or higher, resulted in acute toxicity (cell death). The IC_{50} value of the 120 min-embryos or the 24 h-embryos for S. mirabilis and S. nudus with OPP was around 0.06 mM. In the presence of OPP (0.005-0.05 mM), the proportion (%) of the gastrula keeping the fertilization membrane increased, suggesting that OPP (0.005-0.05 mM) inhibited the hatching process.

M. Wannick *et al.*, 2018 [172] stated that dimethyl fumarate (DMF) is in clinical use for the treatment of psoriasis and multiple sclerosis. The mode of action of DMF in inflammatory skin diseases has remained elusive. Upon ingestion, DMF is converted to monomethyl fumarate that activates the hydroxycarboxylic acid receptor 2 (HCA_2). DMF treatment ameliorated skin lesions in $Hca2^{+/+}$ but not in $Hca2^{-/-}$ animals. These findings demonstrate that HCA_2 is a molecular target of DMF treatment in EBA and suggest that HCA_2 activation limits skin pathology by inhibiting the infiltration of neutrophils and monocytes into the skin.

P. Perini *et al.*, 2018 [173] stated that dimethyl fumarate (DMF) is approved as first line therapy for relapsing-remitting multiple sclerosis (RRMS). In some (3%) patients, DMF induces a marked lymphopenia. A 39-year-old female patient with RRMS that developed HSE temporally associated with a marked and sudden drop in lymphocyte count, from 1200/μl to 600/μl, in the peripheral blood. HSE DNA was demonstrated in the cerebrospinal fluid. Three months later the patient presented only a mild short-term memory deficit. Lymphocyte count returned to normal values (1120/μl) after DMF discontinuation. That case points out the necessity of further studies on DMF-related lymphopenia.

J. L. Andersen *et al.*, 2018 [174] stated that dimethyl fumarate (DMF) has been applied for decades in the treatment of psoriasis and now also multiple sclerosis. Based on a 1.9 Å resolution crystal structure of the C-terminal kinase domain of the mouse p90 Ribosomal S6 Kinase 2 (RSK2) inhibited by DMF. DMF reacts covalently as a Michael acceptor to a conserved cysteine residue in the αF-helix of RSK/MSKs. Binding of DMF prevents the activation loop of the kinase from engaging substrate, and stabilizes an auto-inhibitory αL-helix, thus pointing to an effective, allosteric mechanism of kinase inhibition. The biochemical and cell biological characteristics of DMF inhibition of RSK/MSKs are consistent with the clinical protocols of DMF treatment.

A. A. Ahmed *et al.*, 2018 [175] examined to investigate the effect of dimethyl fumarate (DMF) on Toll-like receptor in rats. Sixty adult male Sprague-Dawley rats were randomly allocated into three groups. group I: rats received the vehicles only; group II: rats were treated with 5 mg/kg per day for 7 days; and group III: rats were given DMF 25 mg/kg for 28 days. Dimethyl fumarate exerted cardioprotective effect against ISO-induced cardiac hypertrophy. This effect is suggested to be through interfering with TLR signaling pathway.

D. Ontaneda *et al.*, 2018 [176] examined about Previous real-world comparative research of MS disease modifying therapies (DMTs) of dimethyl fumarate (DMF). The objective of the study was to compare the annualized relapse rate (ARR) and risk of relapse patients who have switched from a platform therapy to DMF. The risk of relapse was also significantly lower when switching to DMF than TERI (HR = 0.679 [0.503, 0.917]). In this analysis, the effectiveness profiles for those oral DMT users specifically switching from platform therapies are consistent with findings from previous research conducted among all oral DMT users, regardless of prior therapy.

C. Wang 2019 [177] examined about graphene oxide based molecularly imprinted polymers that were prepared as solid-phase extraction column sorbents for detection of di(2-ethylhexyl) phthalate in water samples. The morphology and composition of synthesized sorbents were characterized by scanning electron microscopy, thermo-gravimetric analysis, Raman spectroscopy, and Fourier-transform infrared spectroscopy. The developed extraction procedures coupled to high-performance liquid chromatography exhibited good linearity (0.2–500 μg/L), low limit of detection (0.052 μg/L), and good precision (RSD<5.7%) under optimized conditions. The developed extraction technique has been successfully applied for extracting di(2-ethylhexyl) phthalate.

Y. Zhang *et al.*, 2019 [178] examined about mobile smart phones. They examined the data up till the year 2017. According to their study almost 30% of the whole word population uses mobile smart phones. Recycling process for mobile smart phones is not very common around the world and most of the mobile smart phones are discarded as e-waste, that contribute to the probability of the chemicals they contain to release into the environment. Present study was based on the presence and concentration measurements of 52 main flame retardants (FRs) synthetic chemical agents. These flame retardants (FRs) were extracted from 8 components of seven different models of largely manufactured mobile smart phones. It was identified first time that mobile phones can have these toxicants.

M. Brits *et al.*, 2019 [179] examined about Flame retardants (FRs). Their work was about organophosphorus flame retardants (OPFRs) and brominated flame retardants (BFRs). In this study concentrations of eight polybrominated diphenyl ethers (PBDEs), two alternative BFRs and ten OPFRs were determined in indoor dust ($n = 20$) and pet cat hair ($n = 11$) from South Africa. The OPFRs were the major FRs, contributing to more than 97% of the total FR concentration. The results indicate that OPFRs are ubiquitous in South African indoor environment. Indoor dust is a major source of human exposure to environmental contaminants. This can for example occur through hand-to-mouth contact of toddlers. The presence of FRs, in particular high concentrations of OPFRs, suggests that children and indoor pet cats may have greater exposure to FRs than adults.

C. F. Estill *et al.*, 2019 [180] examined about chlorinated alkyl and non-chlorinated aryl organophosphate flame retardants (OPFRs) and some brominated flame retardants (FR). Organ phosphorous (OP) insecticides are mainly used in agricultural settings. Urinary metabolites of FRs and OPs are known exposure biomarkers to FRs and OP insecticides, respectively. An analytical approach was devloped to quantify in 0.2 mL urine 10 FRs and six OP insecticide metabolites. The method relies on enzymatic deconjugation, isotope dilution tandem mass spectrometry and high-performance liquid chromatography. Detection limit was 0.05 to 0.5 ng mL^{-1}, accuracy 89 to 118%, and precision was <10%.

Y.-J. Hu *et al.*, 2019 [181] examined about inhalation exposure to flame retardants used as additives in human daily activities. They accessed gaseous halogenated flame retardants alternative halogenated flame retardants

(AHFRs), polybrominated diphenyl ethers (PBDEs) and organophosphate esters (OPEs) at indoor and outdoor environments (school, office, and residence) in three districts of a megacity (Guangzhou, China). Results demonstrated that OPEs were the dominant components among all targets. Indoor daily intakes of PBDEs and OPEs were 13-16 times greater than outdoor levels for all age groups.

G. Rivière *et al.*, 2019 [182] examined about brominated flame retardants (BFRs) and perfluoroalkyl acids (PFAAs). These flame retardants are frequently used in human food products. Around 200 baby and common food composite samples have been analyzed for PFAAS, hexabromocyclododecanes, polybrominated biphenyls, polybrominated diphenyl ethers and tetrabromobisphenol A. The dietary exposure of 705 children aged 1-36 months was assessed. PFAAS were detected only in one fish sample. Detection rates varied from 4 to 93% for BFRs. Regarding the provisional health-based guidance values set by EFSA in 2018 for PFOA and PFOS at 0.8 and 1.8 ng kg respectively, 20-100% of children exceeded them.

M. J. Piazza and A. A. Urbanetz, 2019 [183] examined about this review aimed to look into agents and mechanisms characterized as endocrine disrupting chemicals (EDCs). These agents are known to cause several harmful effects to the reproductive system of women and wildlife. There is a wide range of chemicals, developed for commercial use mainly in agriculture, which may cause endocrine disruption. Numerous studies show evidence of environmental contamination. That study was base on analysis and description of the mode of action and the impacts of different EDCs (bisphenols, phthalates, atrazine, polychlorinated and polybrominated biphenyls, DDT-dichlorodiphenyltrichloroethane; DDE-dichlorodiphenyl - dichloroethylene; DDD - dichlorodiphenyl - dichloroethane; and DES-diethylstilbestrol) on the genital area, ovarian steroidogenesis, polycystic ovary syndrome, endometriosis, the structure of the uterus and the vagina, and on the formation of leiomyomas.

Y. Liang *et al.*, 2018 [184] examined about the emission of semi volatile organic compounds (SVOCs). Their work was on the gas-phase diffusion and external mass transfer of SVOC emission process. For the rapid examination of organophosphate flame retardants (OPRFs) they established a new material-air-material (M-A-M) method. The emission test was conducted in 44 mL microchambers They targeted OPFRs, including tris(1-chloro-2-propyl) phosphate (CASRN: 13674-84-5), tris(1,3-dichloro-2-propyl) phosphate (CASRN: 13674-87-8) and tri s(2-chloroethyl) phosphate (CASRN: 115-96-8).

A. M. Al-Salem *et al.*, 2019 [185] examined about tricresyl phosphate (TCP) the organophosphorus flame retardants (OPFRs) used as plasticizer in consumer products. No studies have been attempted so far on TCP to induce hepatotoxicity in human cells. This study confirms the hepatotoxic potential and activation of cancer pathways in TCP treated human hepatocellular cells (HepG2). MTT and NRU data showed 39.3% and 49.85% decline in HepG2 survival when exposed to the highest concentration of TCP (400 μM) for 3 days. Comet assay showed 27.1-fold greater DNA damage in cells treated with TCP (400 μM).

H. Kang *et al.*, 2019 [186] examined about organophosphate esters (OPEs). In vitro studies have suggested adverse effects of OPEs on the kidney. Despite accumulating evidence indicating widespread exposure to OPEs, their association with renal function in humans has not been evaluated. This study aims to investigate the association between OPEs exposure and chronic kidney disease (CKD) among the general US adult population by employing data from the US National Health and Nutrition Examination Survey

(NHANES) 2013-2014. Among 1578 adults who were not currently pregnant, the associations of the urinary OPE metabolites with two CKD-related parameters, i.e., estimated glomerular filtration rate (eGFR) and albumin-to-creatinine ratio (ACR), and CKD were assessed. Experimental studies are warranted to understand the potential mechanisms underlying this observation.

S. Yuan *et al.*, 2019 [187] examined about aryl phosphorus-containing flame retardants (aryl-PFRs). In this study, the cytotoxicity, DNA damage, mitochondrial impairment and the involved molecular mechanisms of certain frequently detectable aryl-PFRs, including methyl diphenyl phosphate (MDPP), 2-ethylhexyldiphenyl phosphate (EHDPP), isodecyl diphenyl phosphate (IDPP), cresyl diphenyl phosphate (CDP), bisphenol-A bis (diphenyl phosphate) (BDP), and the structurally similar and widely used organophosphorus pesticide chlorpyrifos (CPF), were evaluated in A549 cells using high-content screening (HCS) system. Aryl-PFRs showed different lethal concentration 50 (LC50) values ranging from 97.94 to 546.85 μM in A549 cells using CCK-8 assay. This study improves the understanding of PFRs-induced adverse outcomes and the involved molecular mechanism.

Y. Q. Ren *et al.*, 2019 [188] examined TSP and 9-stage size-segregated samples that were analyzed for organic aerosols (OA) on a molecular level. N-Alkanes were the dominant compound class during the whole campaign, followed by fatty acids. High molecular weight (HMW) n-alkanes and fatty acids dominated in the coarse mode particles (>1.1 μm) during the dust event, indicating they were mostly originated from surface soil and plants in the upwind regions. Low-volatile anthropogenic compounds such as benzo(e)pyrene (BeP) and bisphenol A (BPA) dominated in the fine mode particles. This work suggests an important role of multiphase reaction in evolution of aerosol chemistry during the dust long-range transport process.

R. M. Martinez *et al.*, 2019 [189] examined and found that phenols and phthalates are potential endocrine disrupting chemicals (EDCs). These EDCs dysregulate a number of biomolecules and pathways, including microRNAs. MicroRNAs can be carried in transport systems called extracellular vesicle (EV) that are present in most biofluids. These findings provide insight regarding the molecular mechanisms underlying adverse effects of phenol and phthalate exposure on female fertility.

R. M. Shaffer *et al.*, 2019 [190] examined and find that phthalates are association with gestational diabetes (GDM). They evaluated the association between 11 urinary phthalate metabolites. They used multi-variable logistic regression to examine the odds of GDM and IGT in relation to an interquartile-range (IQR) increase in natural log (ln)-transformed, specific gravity (SG)-adjusted first trimester (T1) and average of T1 and third trimester (T3) ("T1T3avg") phthalate metabolite concentrations. In secondary analyses, most other phthalates were not found to be related to study outcomes. Sensitivity analyses indicated possible strong race-specific associations in Asians.

M. Long *et al.*, 2019 [191] has indicated that some non-inherited factors such as exposure to environmental pollutants are associated with neurodevelopment disorders like autism spectrum disorder (ASD). Studies report that endocrine disrupting compounds (EDCs), including polychlorinated biphenyls, organochlorine pesticides, perfluoroalkyl substances (PFAS), and some metals, have adverse effects on the fetal neurodevelopment. The aim of this study was to measure the amniotic fluid (AF) levels of EDCs and metals as well as the receptor transactivities induced by AF and investigate the possible link between prenatal exposure to EDCs and heavy metals and ASD risk.

D. P. Bockmühl *et al.*, 2019 [192] illustrated that elimination of microbial contaminations from textile is an important aspect of laundering textiles. Although the framework for institutional laundering is well regulated to ensure hygienic cleanliness via the use of e.g. high temperatures and bleaching agents. In both cases, energy efficiency of appliances is a main driver, in a general decrease in washing temperatures which in turn can impact the antimicrobial efficacy of laundering. Thus, the different factors influencing the input and removal of microbial cells in the laundering process and possible adverse effects of microbial contaminants in the washing machine and on the textiles as well as suitable counteractions are discussed in this article, focusing on the clinical area but also considering the domestic environment, which will gain importance in the future, e.g. by the increase of elderly and ill persons being cared for at home.

A. de la Torre *et al.*, 2020 [193] examined about 52 pollutants including organophosphate flame retardants and plasticizers (OPs), novel brominated flame retardants (NBFRs) and polybrominated diphenyl ethers (PBDEs) were evaluated in household dust particles from Belgium, Italy and Spain. Pollutant pattern was dominated by ∑OPs (12.8 µg/g; median) followed in decreasing order by ∑PBDEs (229 ng/g), deca bromodiphenyl ethane (DBDPE, 130 ng/g), 1.35 ng/g), hexabromobenzene (HBB, 0.28 ng/g), 1,2-bis(2,4,6-tribromophenoxy)ethane (BTBPE) and finally pentabromoethylbenzene (PBEB, 0.03 ng/g). Country differences and substitution of regulated chemicals by unregulated ones were explored. Calculated EDI levels even at worst case scenario were below available reference dose (RFD) values in all cases.

N. Wemken *et al.*, 2020 [194] examined about the brominated flame retardants (BFRs). They examined the eight polybrominated diphenyl ethers (PBDEs), decabromodiphenyl ethane (DBDPE) and hexabromocyclododecane (HBCDD) in 16 pools of human milk from Ireland. Concentrations of BDEs-47, -99, -100, -153, and HBCDD were significantly lower ($p < 0.05$) than those in Irish human milk collected in 2011. This suggests increased use of DBDPE and Octa-BDE. This suggests exposure pathways not included in the model like dermal uptake from fabrics may be important, and highlights knowledge gaps about the human half-lives and bioavailability of these contaminants.

CHAPTER # 3

EXPERIMENTAL

3.0 Experimental work:

3.1 Solvents and chemicals;

General purpose grade organic solvents and chemical agents were obtained from available suppliers in the market. Tetrahydrofuran (THF), n-Hexane, Toluene, Acetonitrile, Acetone, Potassium hydroxide and Acetic anhydride (Merck) were purchased from New Chemical Company, glacial acetic acid, Ethyl acetate, HPLC grade methanol and Cyclohexane General purpose grade were purchased from supplier (lab line) in Lahore city, Pakistan. Sodium sulfate (anhydrous) from Dae-Jung (Korea) Grade three water was produced by Milli Q instrument. Flame retardants (listed in table 3.1), phthalates standards (certified reference materials, CRMs, and NIST standard reference material, CRM-EP002, CRM-PV001, SRM 3074, listed in table 3.2), d10-anthracene, MIT, CIT, OIT, Triclosan, 4-dimethylaminopyridine (DMAP), Ortho-phenylphenol (OPP) and Dimethyl fumarate were obtained from chem-service; local supplier of UK (greyhound).

Table 3.1 Abbreviated names of flame retardants:

Sr. No.	Abbreviated name	Full name
1	BBMP	2,2-bis(bromomethyl)-1,3-propanediol
2	TBP	Tributyl Phosphate
3	TiBP	Triisobutyl Phosphate
4	BDBPP	Bis-(2,3-dibromopropyl) Phosphate
5	PBEB	Pentabromoethylbenzene
6	TDCPP	Tris(1,3-dichloro-2-propyl) Phosphate
7	TBEP	Tris(2-butoxyethyl) Phosphate

8	TCPP	Tris(2-chloropropyl) Phosphate
9	Penta-BBs	2,2′,4,5,6-Tetrabromobiphenyl
10	Tetra-BDEs	Tetrabromodiphenyl ether
11	TBECH	Tetrabromoethylcyclohexane
12	TOCP	Tri-o-cresyl Phosphate
13	MDPP	4-(tert-butyl)phenyl diphenyl Phosphate
14	TMCP	Tri-m-cresyl Phosphate
15	Penta-BDEs	Pentabromodiphenyl ether
16	TPP	Triphenyl Phosphate
17	TEHP	Tris(2-ethylhexyl) Phosphate
18	TBBPA	Tetrabromobisphenol A
19	DBPP	Bis(tert-butylphenyl) phenyl Phosphate
20	TDBPP	Tris-(2,3-dibromopropyl) Phosphate
21	V6	2,2-Bis(chloromethyl)trimethylene bis(bis(2-cloroethyl) phosphate)
22	TPCP	Tri-p-cresyl Phosphate
23	TBB	2-Ethylhexyl tetrabromobenzoate
24	Hexa-BBs	2,2′,4,4′,5,5′-hexabromobiphenyl
25	DBHC-TCTD	5,6-dibromo-1,10,11,12,13,13-hexachloro-11-tricyclo[8.2.1.02,9]tridecane
26	TBPH	Bis(2-ethylhexoyl)-2,3,4,5-tetrabromophthalate
27	Hexa-BDEs	Hexabromodiphenyl ether
28	Bis-TBBPA	Bis-(2,3-dibromopropyl) ether
29	BTBPE	1,2-bis(2,4,6-tribromphenoxy) ethane

30	TBPP	Tris-(4-tert-butylphenyl) Phosphate
31	TXP	Trixylyl Phosphate
32	Hepta-BDEs	Heptabromodiphenyl ether
33	Octa-BBs	Octabromobiphenyl
34	Octa-BDEs	Octabromodiphenyl ether
35	Deca-BB	Decabromobiphenyl
36	HBCDD	Hexabromocyclododecane
37	Dechlorane-Plus	Bis(hexachlorocyclopentadieno)cyclooctane
38	TEPA	Tris-(aziridinyl)phosphine oxide
39	TCEP	Tris(2-chloroethyl) Phosphate
40	TMP	Trimethyl Phosphate
41	TEP	Triethyl Phosphate
42	Deca-BDEs	Decabromodiphenyl ether
43	DBDPE	Decabromodiphenylethane

Table 3.2 Chemical Abstracts Service (CAS) numbers of phthalate contents.

Sr. No.	Phthalate content full name	Phthalate content short name	CAS-No.
1	Diisononyl phthalate	DINP	28553-12-0/ 68515-48-0
2	Diisodecyl phthalate	DIDP	26761-40-0/ 68515-49-1
3	Di (heptyl, nonyl, undecyl) phthalate	DHNUP	68515-42-4
4	Diundecyl phthalate	DUDP	3648-20-2
5	Di-n-hexyl phthalate	DnHP	84-75-3
6	Dihexyl phthalate	DHP	68515-50-4
7	Diisopentyl phthalate	DIPP	605-50-5
8	n-Pentyl-isopentyl phthalate	nPIPP	776297-69-9
9	Dicyclohexyl phthalate	DCHP	84-61-7
10	Diisooctyl phthalate	DIOP	27554-26-3
11	Di-n-pentyl phthalate	DnPP	131-18-0
12	Diisohexyl phthalate	DIHP	71850-09-4
13	Benzyl butyl phthalate	BBP	85-68-7

14	Diisoheptyl phthalate	DIHpP	71888-89-6
15	Di-2-Ethyl Hexyl phthalate	DEHP	117-81-7
16	Di-n-heptyl phthalate	DnHpP	3648-21-3
17	Dibenzyl phthalate	DBzP	523-31-9
18	Di-n-Octyl phthalate	DnOP	117-84-0
19	Di-(2-Propyl Heptyl) phthalate	DPHP	53306-50-0
20	Dinonyl phthalate	DNP	84-76-4
21	Dimethyl phthalate	DMP	131-11-3
22	Diethyl phthalate	DEP	84-66-2
23	Dipropyl phthalate	DPrP	131-16-8
24	Diiso butyl phthalate	DIBP	84-69-5
26	Diphenyl phthalate	DPhP	84-62-8
25	Bis(2-methoxyethyl) phthalate	DMEP	117-82-8
27	Di-butyl phthalate	DBP	84-74-2
28	Dodecyl phthalate	DDP	84-77-5

3.2 Preparation of standard, calibration standard & stock solutions:

3.2.1 For flame-retardants:

On an analytical balance Model: ML 204/01 Mattler Toledo each flame-retardant were measured 50 + 0.1 mg in a volumetric flask of 10 mL capacity and volumes were filled up to the mark with pure toluene. That prepared solutions were marked as 5000 mg L^{-1} standard stock solution. A working standard solution of 100 mg L^{-1} was prepared from that stock solutions in a separate 10 mL measuring flask and at 4 °C put in refrigerator. Six calibration standard solution points of 0.25 mg L^{-1}, 0.5 mg L^{-1}, 1.0 mg L^{-1}, 2.0 mg L^{-1}, 3.0 mg L^{-1} and 5.0 mg L^{-1} were prepared. These were for all flame retardants in combination. These were prepared from 100 mg L^{-1} standard operating solutions. Volume taken for each calibration standard solution point from 100 mg L^{-1} standard solutions in 10 mL volumetric bottles was 25 μL, 50 μL, 100 μL, 200 μL, 300 μL and 500 μL.

3.2.2 For phthalate contents:

Initially 1000 mg L^{-1} stock solution of anthracene-d10 internal standard (IS) was prepared. 100 mg of anthracene-d10 was dissolved in n-hexane. It was prepared in volumetric bottle of 100 mL capacity. For the preparation of 2 mg

L^{-1} working internal standard solution 1000 mg L^{-1} stock solution was diluted in n-hexane.

Phthalates standard stock solutions of 1000 mg L^{-1} were prepared in n-hexane. Phthalate standards of 100 mg of each was weight in a volumetric flask of 100 mL and volume markup were made in n-hexane. For phthalates working standard solutions of 200 mg L^{-1} were prepared from 1000 mg L^{-1} stock solutions. 5.0 mL of 1000 mg L^{-1} stock solutions were taken into 25 volumetric flask and diluted up to the mark with n -hexane. Six calibration standard solution points of 0.2 mg L^{-1}, 0.9 mg L^{-1}, 2.0 mg L^{-1}, 3.5 mg L^{-1}, 5.0 mg L^{-1} and 10.0 mg L^{-1} were prepared. These were for all phthalate contents in combination. These were prepared from 200 mg L^{-1} standard operating solutions. Volume taken for each calibration standard solution points from 200 mg L^{-1} standard solutions in 20 mL volumetric bottles was 20 μL, 90 μL, 200 μL, 350 μL, 500 μL and 1000 μL.

3.2.3 For biocide contents:

Biocide contents standard stock solutions of 1000 mg L^{-1} were prepared at initial point separately. Each biocide content 10.00 mg was weight sequentially on an analytical balance. Each biocide content was taken in 10 mL separate volumetric bottles and was diluted using methanol as solvent. For biocide contents working standard solutions of 100 mg L^{-1} were prepared separately. 500 μL solution of 1000 mg L^{-1} biocide contents stock solutions were taken in separate 5 mL bottles and were diluted in methanol. Five calibration standard solution points of 1.0 mg L^{-1}, 2.0 mg L^{-1}, 3.0 mg L^{-1}, 4.0 mg L^{-1} and 5.0 mg L^{-1} were prepared in combination for biocide contents from working standard solutions of 100 mg L^{-1}.

3.2.4 For Ortho-phenylphenol (OPP):

Potassium hydroxide 1 mol L^{-1} was prepared by dissolving 56.11 g in 1000 mL of grade three water. 0.1 g 4-dimethylaminopyridine (DMAP) was weighed and dissolved in 5mL of grade three water. Ortho-phenylphenol (OPP) (purity ≥ 98%) was weighed 0.01 g in 10 mL measuring flask. Solution was diluted using acetone up to specified mark to obtain primary stock solution (1000 mg L^{-1}). Dilution was made of that primary stock solution from 1.0 mL to 10 mL in a measuring flask with acetone for acquiring working stock solution (100 mg L^{-1}). Appropriate dilutions were made from 100 mg L^{-1} working stock solution for the calibration curve points (i.e. 0.1 mg L^{-1}, 0.2 mg L^{-1}, 0.5 mg L^{-1}, 1.0 mg L^{-1}, 2.0 mg L^{-1}, and 5.0 mg L^{-1}). Acetylation step for each curve point was performed in 40 mL screw capped vial containing 30 mL of potassium hydroxide 1mol L^{-1} solution. Anthracene d-10 was weighed 0.01 g in 50 mL measuring flask. Solution was diluted using acetone for acquiring 200 mg L^{-1} working internal standard solution.

3.2.5 For Dimethyl fumarate (DMFu)

1000 mg L^{-1} working standard solution of dimethyl fumarate (DMFu) was prepared by dissolving 50.0 ± 0.5 mg in ethyl acetate in 50 mL volumetric flask. Intermediate standard solution of 10 mg L^{-1} was prepared from working standard solution. Calibration standard solutions of 0.01 mg L^{-1}, 0.025 mg L^{-1}, 0.05 mg L^{-1}, 0.10 mg L^{-1}, 0.25 mg L^{-1} and 0.50 mg L^{-1} were prepared from working standard solution of 10 mg L^{-1}.

3.2.6 For chrome (VI)

Primary stock solution of chrome (VI) 1000 mgL^{-1} was prepared by dissolving 2.829 g of potassium dichromate $K_2Cr_2O_7$ in grade three water. That solution was prepared in the 1000 mL volumetric flask in grade 3 water. Working standard solution 1.0 mgL^{-1} was prepared by taking 0.1 mL of primary stock solution in 100 mL volumetric flask. This working standard solution was used to prepare calibration points (0.002 mg L^{-1}, 0.005 mg L^{-1}, 0.01 mg L^{-1}, 0.02 mg L^{-1}, 0.06 mg L^{-1}, 0.1 mg L^{-1}, 0.16 mg L^{-1}, 0.2 mg L^{-1}, 0.3 mg L^{-1}, and 0.4 mgL^{-1})

3.3 Collection of test specimens for experimental examination:

Test specimens were collected for experimental examination from the textile and leather manufacturing industries and retailer shops in the Lahore city of Pakistan. Finished textile and leather articles were collected from retailer shops and grocery stores. Finished garments, fabric products, non-printed and printed fabrics were collected from the textile and leather manufacturing industries between the period January and June 2019.

3.4 Test specimen's preparation:

3.4.1 For flame retardants:

Test specimen was randomly selected for experimental examination. If the test specimen was of the same color, it was grinded or sliced into small pieces about 1 mm × 1 mm size. The test specimen was selected as a color percentage and cut into smaller pieces about 1 mm × 1 mm, when the test sample was patterned or multi-colored. The test sample was selected from the uncovered parts, if the removable coating of paint of any other substance was present on the template. The cover or coating was removed with a possible source as surgical blade etc. The test specimens that were collected from market were examined for target analytes and some test specimens from them were selected as blank matrix. In these test specimens' pure standards were spiked for examining different parameters. Various analysis parameters as adequate time for analysis, methods suitability for test specimens, digestion time duration etc. were optimized. Adequate solvent was found toluene appropriate time was selected 60 min and excellent recovery method was selected ultrasonic. (figure "4.1.1" and table 4.1.1, 4.1.2 and 4.1.3 can be explored).

3.4.2 For phthalates:

From various parts of the consumer's product the test sample was taken randomly. When the test specimen was child toy the power sources as batteries were removed. These parts were discarded after separation from the test specimen. So, they were not considered the parts of the subjected test specimen. The test sample was shopped or grinded into small pieces and blended well if that was the same colored without any color pattern. In the case of a multi-colored test sample or pattern the test sample was collected according to color pattern or color proportion. Possibly all respected colors were covered in test specimen sampling. After sampling they were grinded or cut into small pieces and mixed to make test specimen homogeneous.

1 mg L^{-1}, 2 mg L^{-1} and 5 mg L^{-1} pure standards were spiked in three test specimens A, B and C. These test specimens were used as blank matrix in the subjected study. Different studies were performed on these test specimens. Suitability of extraction method, time and extraction solution were examined. Adequate solvent was found tetrahydrofuran appropriate time was selected 60 min and excellent recovery method was selected ultrasonic. (figure "4.2.1" and table 4.2.1, 4.2.2 and 4.2.3 can be explored).

3.4.3 Biocides contents:

For biocides examination the test sample was randomly taken from various parts of the test sample from textile consumer articles. If the test specimen was of the same color, it was grinded or sliced into small pieces about 5 mm × 5 mm size. The test specimen was selected as a color percentage and cut into smaller pieces about 5 mm × 5 mm, when the test sample was patterned or multi-colored.

The test specimens that were collected from market were examined for target analytes and some test specimens from them were selected as blank matrix. In these test specimen's pure standards was spiked 3.0 mg L^{-1} for examining different parameters. Various analysis parameters as adequate time for analysis, methods suitability for test specimens, digestion time duration etc. were optimized. Adequate solvent was found methanol, appropriate time was selected 30 min and excellent recovery method was selected ultrasonic. (figure "4.3.1" and table 4.3.1, 4.3.2 and 4.3.3 can be explored).

3.4.4 For Ortho-phenylphenol (OPP):

Test specimen was randomly selected for experimental examination. If the test specimen was of the same color, it was grinded or sliced into small pieces about 5 mm × 5 mm size. The test specimen was selected as a color percentage and cut into smaller pieces about 5 mm × 5 mm, when the test sample was patterned or multi-colored. The test sample was selected from the uncovered parts, if the removable coating of paint of any other substance was present on the template. The cover or coating was removed with a possible source as surgical blade etc. The test specimens that were collected from market were examined for target analytes and some test specimens from them were selected as blank matrix. In these test specimens' pure standards were spiked for examining different parameters. Various analysis parameters as adequate time for analysis, methods suitability for test specimens, digestion time duration etc. were optimized. Adequate solvent was found 0.1 mg L^{-1} KOH, appropriate time was selected 60 min and excellent recovery method was selected ultrasonic. (results were summarized in figure "4.4.2" and table 4.4.2, 4.4.3 and 4.4.4 in additional data).

3.4.5 For Dimethyl fumarate (DMFu):

For rubber, PU fabric, leather samples and other test specimens of textile related consumer articles, the test specimen was cute into approximately 5mm × 5mm pieces, while for silica gel specimens, the sample was grinded well to make a powder.

A, B and C these three specimens were chosen as blank matrix for further proceedings. These three test specimens were utilized to spike pure reference standards and were examined by using multiple types of solvents, extraction methods and time. Ultrasonic assisted extraction method, ethyl acetate organic solvent, 40 min extraction time shows finest recovery results. Hence ethyl acetate was selected solvent for extraction, 40 min time period for extraction and ultrasonic assisted extraction. (Results were presented in figure 4.5.2 and table 4.5.0, 4.5.1, 4.5.2 in results and discussion portion).

3.4.6 For Chrome (VI): Chrome (VI) with aging

Representative specimen was selected from multiple parts of the specimens under examination, cut into pieces of 3mm × 3mm and weighted accurately 5gm, accurately weight 2±0.01gm from this for proceeding analysis and 3±0.001gm for volatile matter. 2gm test specimen was taken in a petri plate and simulated the aging effect of (sea) shipment conditions of Cr (VI) development in leather test specimens by an accelerated aging condition using an oven-controlled humidity ≤5% at 80±2°C for 24 hours ±5 minutes. Petri dish was taken out from oven after incubation time and allowed to cool in a desiccator for 30 minutes. After desiccation extraction was performed according to the procedure (Cr (VI) extraction) described below.

3.4.7 Chrome (VI) without aging:

Representative specimen was selected from multiple parts of the specimens under examination, cut into pieces of 3mm × 3mm and weighted 5 gm, accurately weight 2±0.01gm from this for proceeding analysis (Cr (VI) extraction) and 3±0.001gm for volatile matter.

3.5 Test specimen's extraction:

3.5.1 For flame retardants:

In a glass culture tube of 10 mL volume capacity the test specimen was weighed accurately of 0.3 g ± 0.01 g. In the glass culture tube 3 mL toluene was added. For the extraction of flame retardants, the glass culture tube was capped tightly and sonicated for 60±2 min at temperature 60±2 °C. No further extraction was needed as 60 min extraction process was enough for whole extraction of flame retardants. The subjected tube was cooled for 2 min in ice bath to room temperature, vigorously shacked on vortex shaker and for 30 min kept aside on separate place. In case, if the solution was clear. It was filtered using 0.2 µm glass wool filter and directly taken for instrumental analysis. Hazy or turbid solution was centrifuged at 6000 rpm for at least 10 min, filtered with 0.2 µm glass wool filter. Then for subsequent analysis it was filled in a GC vial of 1.5 mL volume capacity.

3.5.2 For phthalates:

In a glass culture tube of 10 mL volume capacity the test specimen was weighed accurately of 0.1 g ± 0.01 g. In that glass culture tube containing test specimen,

5 mL tetrahydrofuran was added. For the extraction of phthalate contents, the glass culture tube was capped tightly and sonicated for 60±2 min at temperature 60±2 °C. No further extraction was needed as 60 min extraction process was enough for whole extraction of the phthalate contents. The subjected tube was cooled for 2 min in ice bath to room temperature. Dropwise internal standard solution (2.0 mg L^{-1} concentration of d10-anthracene) 5.0 mL was added to the solution. The final concentration of internal standard was 1 mg L^{-1} in the solution. The glass culture tube was vigorously shacked on vortex shaker and for 30 min kept aside on separate place. In case, if the solution was clear. It was filtered using 0.45 µm glass wool filter and directly taken for instrumental analysis. Hazy or turbid solution was centrifuged at 6000 rpm for at least 10 min, filtered with 0.45 µm glass wool filter. Then for subsequent analysis it was filled in a GC vial of 2.0 mL volume capacity.

3.5.3 For biocides:

In a reagent bottle of 100 mL volume capacity the test specimen was weighed accurately of 5.00 g ± 0.01 g. In that reagent bottle containing test specimen, 80 mL methanol was added. For the extraction of biocide contents, the reagent bottle was capped tightly and sonicated for 30±1 min at temperature 50±2 °C. No further extraction was needed as 30 min extraction process was enough for whole extraction of the biocide contents. The subjected reagent bottle was cooled for 2 min in ice bath to room temperature. The reagent bottle was vigorously shacked on vortex shaker and for 30 min kept aside on separate place. In case, if the solution was clear, its volume was reduced to 2 mL approximately at 55 °C temperature on rotary vacuum evaporator, filtered using 0.45 µm glass wool filter and directly taken for instrumental analysis. Hazy or turbid solution was centrifuged at 6000 rpm for at least 10 min, its volume was reduced to 2 mL approximately at 55 °C temperature on rotary vacuum evaporator, filtered with 0.45 μm glass wool filter. Then for subsequent analysis it was filled in a GC vial of 1.5 mL volume capacity.

3.5.4 For Ortho-phenylphenol (OPP):

Test specimen was weighed 5.00 g accurately using analytical balance Mettler Toledo ML 204/01 in a 100 mL reagent bottle. 1.0 mg L^{-1} potassium hydroxide solution 60 mL was added in the glass bottle, capped and was kept in the sonication instrument for 60 mins at 70 °C.

3.5.4.1 Acetylation step for Ortho-phenylphenol (OPP):

The extract was cooled to room temperature and 30 mL of extract was taken in 40 mL capacity capped vial. DMAP solution 0.5 mL and 2.5 mL acetic anhydride was added to the vial, capped and shake for 5 mins on vortexed shaker followed by orbital shaking for 60 mins at 200 rpm. Then 5 mL n-Hexane was added and vortexed again for 5 mins followed by 30 mins orbital shaking at 200 rpm. 30 mins extraction process was enough for whole extraction of OPP contents and there was no need for further extraction. Extract (Organic n-Hexane layer that contain extracted target analyte) was taken 1mL into a 1.5 mL GC vial, internal standard 10 mg L^{-1} was added and analyzed on GC-MS.

3.5.5 For Dimethyl fumarate (DMFu):

Accurately weighed 1.00 g test specimen was taken into a culture tube. 10 mL ethyl acetate was added to the tube and was capped tightly. Mixture was sonicated at 70 °C for 40 min, filtered with glass wool 0.45 μm syringe filter for GC/MS analysis. 40 min extraction process was enough for whole extraction of DMFu contents and there was no need for further extraction.

3.5.6 For chrome (VI):

Extraction solution 100 mL degassed with nitrogen for 5 minutes was taken in 250mL stoppered conical flask. 2 gm test specimen was extracted by shaking the flask containing test specimen on a mechanical shaker for 3hr±5 minutes at 100±10 minutes^{-1} frequency at room temperature (20-25°C). For UV method after extraction the content of the flask were filtered with Whatman filter paper # 42 into a 250mL conical flask. Extract was taken from the flask and pH was checked. If pH was in the range 7.0 to 8.0, the test specimen was processed and in case pH not in range the procedure was repeated with new test specimen, if still not in range the procedure was repeated with 1.0±0.01gm of test specimen. For UFLC analysis the content of the flask was filtered with 0.45µm PTFE filter in 1.5 mL amber GC Vial.

If the test specimen solution was colorless 20 mL of this was taken in 50 mL volumetric flask and markup with extraction solution to make S_1 solution.

3.5.6.1 Cartridges cleanup:

If the test specimen's extract was not colorless cartridges cleanup was performed with test specimen to remove intense color. Cartridges were pretreated with 5 mL methanol, 5mL grade 3 water, 10 mL extraction solution and finally with 5mL test specimen's extracted solution. Then 50 mL colored test specimen solution was passed quantitatively through cartridges on solid phase extraction system with vacuum device and elute was collected drop by drop in a 30 mL glass vial. Procedure was repeated upto 4 time and maximum 4 cartridges were used for a single test specimen, if still color was present in the eluted solvent.

3.5.6.2 Charcoal color removal:

After extraction in the extraction flask charcoal was added successively as 0.1 g batch of each time and shake well until the extracted test specimen was colorless. Test specimen was filtered with 3 folds of Whatman filter paper # 42 into a 250 mL conical flask.

- 20 mL of that extracted clean solution was measured and added in a 50 mL measuring flask. Markup to the point with extraction solution to make S_1 solution. Take 20 mL S_1 solution in a volumetric flask 25 mL add 0.5 mL DPC solution 0.5 mL phosphoric acid and markup the volumetric flask with extraction solution and shake well.
- Solution was allowed to stand for 15±5 minutes and the absorbance of this solution was measured at 540nm against the blank solution at spectrophotometer.
- Volatile matter of the test specimen was calculated according to ISO-4684 (15).
- Results were calculated from absorbance of the test specimen according to standard methods (1).

3.6 Forced degradation study for biocides:

For CIT, OIT, MIT and Triclosan rapid destruction studies was conducted. The HCl solution (0.1 N) was used for acid degradation of biocide contents at ambient temperature. 5 mL methanolic solution of biocide content was subjected for acidic degradation for 60.0 min time period in 1.5 mL HCl solution (0.1 N). The NaOH solution (0.1 M) was used for alkaline degradation of biocide contents at ambient temperature. 5 mL methanolic solution of biocide content was subjected for alkaline degradation for 60.0 min time period in 1.5 mL NaOH solution (0.1 M). Heat decomposition was performed on the methanolic solution of biocide contents by exposing to high temperature. 5 mL methanolic solution of biocide content was subjected at high temperature of 80

°C for 72 hours. For oxidative stress degradation 5 mL methanolic solution of biocide content was exposed to H_2O_2 (30% v/v) at ambient temperature for 60.0 min time period. 5 mL methanolic solution of biocide content was exposed to UV-light at 320 nm wavelength for 72 hours for photolytic tress degradation.

3.7 GC-MS Equipment:

In the whole experimental examination latest chromatographic equipment of Shimadzu corporation (GC-2010 ultra plus) was used. There was installed injector AOC-20i, auto sampler AOC-20s and mass selective detector QP2010 ultra plus. GC-MS solution software latest version (version # 2.70 Shimadzu corporation), and updated installed internal library (NIST 127 & 147) was utilized in the whole examination. Suitable columns with maximum temperature of 360 °C DB-XLB column (0.25mm × 15m × 0.25μm), DB-5MS (0.25 mm × 30 m × 0.25 μm), Hypersil C18 (250mm×4.6mm×5μm) etc. was used for test specimen's examination. Full scan operation mode (m/z, 35 to 1000) EI ionization at 70eV with electron multiplier setup 1.2kV was utilized for peak identification and quantification purpose.

3.8 GC-MS instrumental parameters:

Perfluorotributylamine (PFTBA, on internally selected m / z 502, 219 and 69) was used in the latest equipment for performing automatic tuning for mass selective detector. For test specimen's injection split-less injection system mode was operated on the chromatographic system. After test specimen's injection to equipment the purge flow for gas was 1.0 mL min^{-1}. The carrier gas used was highly pure helium gas (99.99 % pure). 1.2 mL min^{-1} was the flow rate of carrier gas. Test specimen's injection volume was as low as 1.0 μL. In most of the test specimen's examination injector temperature was 250 °C. It was preset temperature. Interface temperature was varied for different exanimations studies, mostly was set at 320 °C.

3.9 Temperature programs:

3.9.1 For flame retardants analysis:

50 °C was the initial temperature of the column. That temperature was retained for 3.0 minutes after test specimen injection. Solvent cut off time was 2.5 min. After 3.0 min temperature was ramped from 50 to 200 ° C for 4.0 min at raising rate of 30 ° C min^{-1}. Finally, the column temperature was raised at rate 40 ° C min^{-1} from 200 °C to 340 °C for the next 4.0 min time period. Total run time for analysis on instrument was 20.5 min.

3.9.2 For phthalates analysis:

50 °C was the initial temperature of the column. That temperature was retained for 1.0 min after test specimen injection. Solvent cut off time was 6.0 min. After 1.0 min temperature was ramped from 50 to 310 ° C for 10.0 min at raising rate of 30 ° C min^{-1}. Total run time for analysis on instrument was 19.7 min.

3.9.3 Ortho-phenylphenol (OPP):

100 °C was the initial temperature of the column. That temperature was retained for 1.0 minutes after test specimen injection. Solvent cut off time was 0.5 min. After 1.0 min temperature was ramped from 100 to 200 ° C for 2.0 min at raising rate of 20 ° C min^{-1}. Finally, the column temperature was raised at rate 40 ° C min^{-1} from 200 °C to 300 °C for the next 6.0 min time period. Total run time for analysis on instrument was 17.0 min.

3.9.4 Dimethyl fumarate (DMFu) analysis:

50 °C was the initial temperature of the column. That temperature was retained for 1.0 min after test specimen injection. Solvent cut off time was 0.5 min. After 1.0 min temperature was ramped from 50 to 250 ° C for 3.0 min at raising rate of 70 ° C min^{-1}. Total run time for analysis on instrument was 6.68 min.

3.10 LC- DAD Instrumentation for biocides and chrome (VI) analysis:

In the whole experimental examination, qualitative and quantitative study LC chromatographic system (Shimadzu) was used. Liquid chromatograph (LC - 20AT) with auto sampler (SIL-20AHT), UV DAD detector (SPD-M20A), degasser unit (DGU-20A5R), quaternary pump LC 20AT (Serial No. L20114811607) and column oven (CTO-20A). Latest software 5.71 version of shimadzu corporation was installed internally (Lab solutions software 5.71 version).

3.11 LC- DAD Instrumental conditions for biocides:

For experimental examination of biocide contents, the chromatographic column used was C18 Hypersil (250 mm × 4.6 mm × 5 μm). Wavelength was selected 280 nm for biocide contents. Mobile phase used for biocide contents was methanol (50: 50 / v: v) and 0.4% acetic acid. Column temperature was selected 30°C. Injection volume used in the examination was as low as 30 μL. Solvent was operated on isocratic mode at flow rate of 1.0 mL min^{-1}. Total run time for analysis on instrument was 12.0 min.

3.11.1 LC- DAD Instrumental conditions for chrome (VI):

UV-1800 spectrophotometer instrument CAT-No.206-25400-38 Serial No. A11454806002 CD shimadzu (Japan) was used to take absorbance at 540nm of test specimens. Confirmation of the presence of chrome (VI) was performed on UFLC. The chromatographic column 1C-PAK-A HR waters was used (4.6 mm × 250mm) with 65 °C maximum temperature capacity, 30°C temperature was used during analysis. 20mM $(NH_4)_2SO_4$ and 1mM sodium hydroxide buffer was used as mobile phase, 50μL injection volume; 0.9 mL/min flow rate in isocratic mode, 372 nm was the selected wavelength. Total analysis time for examination on UFLC instrument was 6.0 minutes.

3.12 Instrumental analysis of test specimens on GCMS equipment:

GCMS instrument limits were confirmed just before performing the analysis. The GCMS instrument was conditioned according to (section 3.8) instrumental conditions and (section 3.9) temperature program. Using peak area vs concentration of analyte, calibration curve was established for subjected analytes. Linear regression coefficient (r^2) was > 0.9995. Test specimen blank and method blank were prepared and examined on GCMS instrument for the conformation that there should be no any contamination added during test specimen processing and from reagents. Prepared standard solution according to subjected methods were examined at GCMS instrument for the assessment of the sensitivity of the instrument. Experimental recovery was evaluated according to subjected methods by examining calibration standard check solution, laboratory quality control solution and specimen spike solution on instrument. In the range of 85 – 115 % was recovery of these solutions (results were summarized in table 3.3). Testing samples were examined on the instrument after ensuring all above limits. The target analytes were confirmed in the test specimen by comparing the retention time and mass spectrum of test specimen with standard mass spectrum. The retention time of the test specimen components was matched with the standard component's relative retention time. It was within the acceptable range of deviation from standard retention time (± 0.01). If quantitation wavelength response was too high from the range of prepared calibration curve and it was exceeding from the last point range. That was due to high concentration of analyte in the test specimen. In such cases test specimen extract was diluted with respective solvent to attain the required range. Such test specimens were retested or by reducing test specimens' weight (with half weight) either by increasing final volume of respective solvent (dilution was made in respective solvent).

3.13 Instrumental analysis of test specimens on LC/DAD equipment:

GCMS instrument limits were confirmed just before performing the analysis. The LC/DAD instrument was conditioned according to (section 3.11) instrumental conditions. Using peak area vs concentration of analyte, calibration curve was established for subjected analytes. Linear regression coefficient (r^2) was > 0.995. Test specimen blank and method blank were prepared and examined on GCMS instrument for the conformation of there should be no any contamination added during test specimen processing and from reagents. Prepared standard solution according to subjected methods were examined at GCMS instrument for the assessment of the sensitivity of the instrument. Experimental recovery was evaluated according to subjected methods by examining calibration standard check solution, laboratory quality control solution and specimen spike solution on instrument. In the range of 80 – 120 % was recovery of these solutions (results were summarized in table 3.3). Testing samples were examined on the instrument after ensuring all above limits. The target analytes were confirmed in the test specimen by comparing the retention time and mass spectrum of test specimen with standard mass spectrum. The retention time and UV spectrum of the test specimen components was matched with the standard component's relative retention time and UV spectrum. It was within the acceptable range of deviation from standard retention time (± 0.01) and UV spectrum peak maxima/ minima was within ± 1 nm of reference spectrum. If quantitation wavelength response was too high from the range of prepared calibration curve and it was exceeding from the last point range. That was due to high concentration of analyte in the test specimen. In such cases test specimen extract was diluted with respective solvent to attain the required range. Such test specimens were retested or by reducing test specimens' weight (with half weight) either by increasing final volume of respective solvent (dilution was made in respective solvent).

Table 3.3. Interim parameters utilized during testing:

Parameters name	Method blank	Specimen blank	Sensitivity check	Lab quality check	Specimen spike	Calibration standard check
Short name	MB	SB	SC	LQC	SS	CSC
For flame retardants Spiked (mg L^{-1})	--	--	0.25	3.0	3.0	3.0
Detected (mg L^{-1})*	BDL	BDL	0.24	2.93	2.89	3.02
Standard Deviation	0	0	0.01	0.05	0.08	0.01
% Relative Standard Deviation	0	0	2.89	1.67	2.64	0.47
For Phthalate contents Spiked (mg L^{-1})	--	--	0.20	5.0	5.0	3.0
Detected (mg L^{-1})*	BDL	BDL	0.21	4.95	4.98	3.01
Standard Deviation	0	0	0.01	0.04	0.01	0.01
% Relative Standard Deviation	0	0	3.45	0.71	0.28	0.24
For ortho-phenylphenol Spiked (mg L^{-1})	--	--	0.10	1.0	1.0	1.0
Detected (mg L^{-1})*	BDL	BDL	0.11	0.95	0.98	1.01
Standard Deviation	0	0	0.01	0.04	0.01	0.01
% Relative Standard Deviation	0	0	6.73	3.63	1.43	0.7
For dimethyl fumarate Spiked (mg L^{-1})	--	--	0.10	0.50	0.50	0.50
Detected (mg L^{-1})*	BDL	BDL	0.11	0.49	0.48	0.51

Standard Deviation	0	0	0.01	0.01	0.01	0.01
% Relative Standard Deviation	0	0	6.73	1.43	2.89	1.4
For biocide contents Spiked (mg L^{-1})	--	--	1.0	3.0	3.0	3.0
Detected (mg L^{-1})*	BDL	BDL	1.03	2.95	2.98	3.01
Standard Deviation	0	0	0.02	0.04	0.01	0.01
% Relative Standard Deviation	0	0	2.09	1.19	0.47	0.24

BDL: Below detection limit, *: Three replicate measurements

CHAPTER # 4

RESULT & DISCUSSION

4.0 Results and discussion:

4.1 Flame retardants:

It was found from the results that peak resolution was in better form in test specimens. All the subjected analytes show their peaks within 20.5 min. Their peaks were separated from each other and easily identifiable (figure "4.1.0" identifies results).

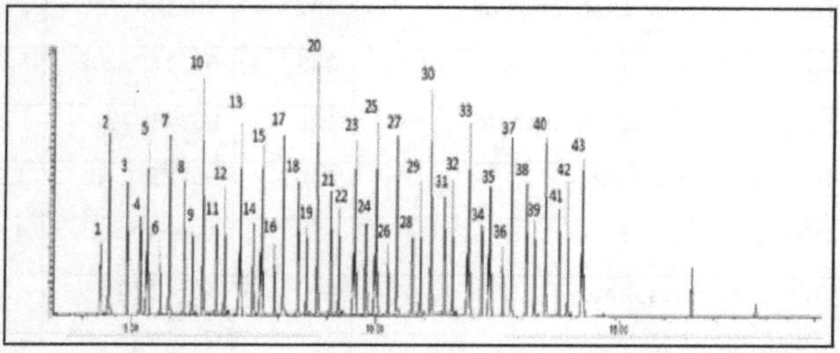

Figure 4.1.0 Reference retention times for flame retardants from a spiked test specimen (3.0 mg L^{-1}), Table 4.1.0 can be explored for 1-43 peaks identities, and section 3.7, 3.8 and 3.9.1 for analysis parameters.

Table 4.1.0 can be screened for mass fragments selected during time of analysis. Quantification of analytes from test specimens were performed on the basis of external standard method. Structure and identity of each detected flame-retardant was matched with the mass spectrum of standards mass spectrum (Figure "1.1 to 1.43" can be explored for result interpretation from annexure 1).

Table 4.1.0 Flame retardants and their mass fragment ions selected for analysis on GC-MS equipment.

Sr. No.	Abbreviated name	Full name	Target ion m/z on GC-MS	Qualifier ions m/z on GC-MS	Retention time (min)
1	TBP	Tributyl Phosphate	99	155, 125	5.40
2	TEPA	Tris-(aziridinyl)phosphine oxide	131	90, 145	5.55
3	TCEP	Tris(2-chloroethyl) Phosphate	249	63, 143, 205	5.76
4	TCPP	Tris(2-chloropropyl) Phosphate	125	99, 41, 277	5.88
5	TMP	Trimethyl Phosphate	110	95, 79, 140	4.26
6	TEP	Triethyl Phosphate	155	129, 99, 81	4.35
7	TiBP	Triisobutyl Phosphate	99	155, 139	4.96
8	BBMP	2,2-bis(bromomethyl)-1,3-propanediol	214	135, 151	5.21
9	PBEB	Pentabromoethylbenzene	500	485, 421, 340	7.15
10	TDCPP	Tris(1,3-dichloro-2-propyl) Phosphate	75	99, 191, 381	7.44
11	TBEP	Tris(2-butoxyethyl) Phosphate	57	125, 199, 85	7.58
12	TPP	Triphenyl Phosphate	326	325, 169, 77	7.96
13	Penta-BBs	2,2´,4,5,6-Tetrabromobiphenyl	388	469, 548	6.06
14	Tetra-BDEs	Tetrabromodiphenyl ether	486	326, 328	6.25
15	TBECH	Tetrabromoethylcyclohexane	267	105, 79, 187	6.54
16	BDBPP	Bis-(2,3-dibromopropyl) Phosphate	337	257,119,219	6.97
17	Penta-BDEs	Pentabromodiphenyl ether	563	404, 406	9.14
18	TPCP	Tri-p-cresyl Phosphate	368	107, 91, 65	9.27
19	TBB	2-Ethylhexyl tetrabromobenzoate	421	393, 312, 70	9.56
20	Hexa-BBs	2,2´,4,4´,5,5´-hexabromobiphenyl	627	467, 549	9.83
21	TEHP	Tris(2-ethylhexyl) Phosphate	99	113, 57	8.23
22	TOCP	Tri-o-cresyl Phosphate	368	165, 277, 91	8.23
23	MDPP	4-(tert-butyl)phenyl diphenyl Phosphate	367	77, 368, 382	8.54

24	TMCP	Tri-m-cresyl Phosphate	368	165, 277, 91	8.81
25	TBBPA	Tetrabromobisphenol A	529	543, 293, 450	10.21
26	DBHC-TCTD	5,6-dibromo-1,10,11,12,13,13-hexachloro-11-tricyclo[8.2.1.02,9]tridecane	268	79, 375, 540	9.97
27	DBPP	Bis(tert-butylphenyl) phenyl Phosphate	423	204, 367, 438	10.48
28	TDBPP	Tris-(2,3-dibromopropyl) Phosphate	419	217, 121,619	10.83
29	V6	2,2-Bis(chloromethyl)trimethylene bis(bis(2-cloroethyl) phosphate)	323	63, 503, 547	10.95

The test specimens that were collected from market were examined for target analytes and some test specimens as 1, 2 and 3 from them were selected as blank matrix. In these test specimens' pure standards of flame retardants (0.25 mg L^{-1}, 1.0 mg L^{-1} and 3.0 mg L^{-1}) were spiked for examining different parameters. Various analysis parameters as adequate time for analysis, methods suitability for test specimens and digestion time duration etc. were optimized. Solvents as chloroform, acetone and toluene, extraction methods as reflux, rotary and sonication, times as 30, 40, 50, 60, 70- and 80 minutes were used alternatively to reproduce subjected flame retardants. Adequate solvent was found toluene, time 30 min and extraction method sonication. (figure "4.1.1" and table 4.1.1, 4.1.2, and 4.1.3 can be explored for result interpretation).

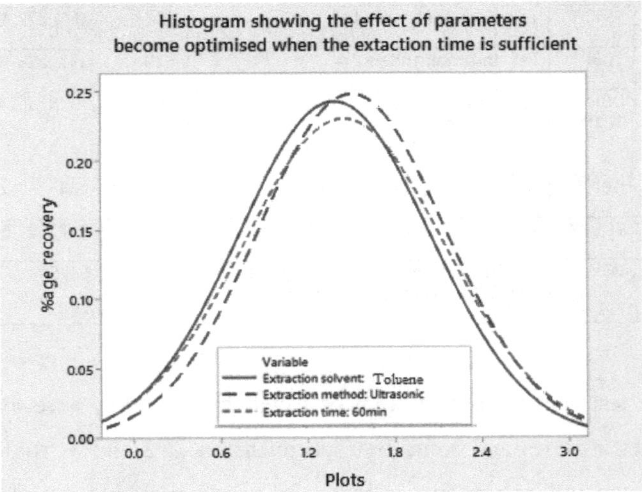

Figure 4.1.1 Results of textile test specimens spiked with flame retardants concentration of 3.0 mg L^{-1} for examining the parameters as extraction solvent, method & time.

Table 4.1.1 Textile test specimens spiked with analyte concentration in mg L^{-1} for examining the extraction effect of different extraction solvents on flame retardants.

Test specimen	Spike Level (mg L^{-1})*	Extraction solvents			SD	%RSD
		Toluene	Chloroform	Acetone		
1	0.25	0.24	0.19	0.20	0.03	13.38
	1.0	0.98	0.65	0.78	0.17	19.66
	3.0	2.93	2.55	2.67	0.21	7.63
2	0.25	0.22	0.17	0.21	0.03	15.55
	1.0	0.94	0.68	0.73	0.16	18.66
	3.0	2.90	2.42	2.59	0.27	9.87
3	0.25	0.26	0.21	0.22	0.02	10.13
	1.0	0.96	0.77	0.88	0.10	11.24
	3.0	2.88	2.72	2.57	0.19	6.71

*= Three replicate measurements, SD =Standard Deviation, %RSD=%Relative Standard Deviation

Table 4.1.2 Textile test specimens spiked with analyte concentration in mg L^{-1} for examining the extraction effect of different extraction methods on flame retardants.

Test specimen	Spike Level (mg L^{-1})*	Extraction methods			SD	%RSD
		Ultrasonic	Reflux	Rotary		
1	0.25	0.24	0.23	0.24	0.01	3.40
	1.0	0.98	0.93	0.78	0.10	10.78
	3.0	2.93	2.85	2.77	0.10	3.44
2	0.25	0.23	0.24	0.21	0.02	7.35
	1.0	0.96	0.94	0.82	0.08	8.33
	3.0	2.95	2.83	2.68	0.14	4.97
3	0.25	0.26	0.22	0.23	0.02	7.61
	1.0	0.95	0.91	0.86	0.06	6.39
	3.0	2.94	2.77	2.72	0.13	4.68

*= Three replicate measurements, SD =Standard Deviation, %RSD=%Relative Standard Deviation

Table 4.1.3 Textile test specimens spiked with analyte concentration in mg L^{-1} for examining the extraction effect of different extraction times on flame retardants.

Test specimen	Spike Level (mg L^{-1})*	Times for extraction						SD	%RSD
		30 min	40 min	50 min	60 min	70 min	80 min		
1	0.25	0.18	0.20	0.23	0.24	0.23	0.24	0.02	6.45
	1.0	0.58	0.64	0.91	0.95	0.98	0.95	0.12	13.01
	3.0	2.52	2.59	2.89	2.95	3.02	2.94	0.14	4.77
2	0.25	0.17	0.19	0.21	0.22	0.23	0.22	0.02	8.70
	1.0	0.56	0.68	0.93	0.99	0.98	0.95	0.12	12.55
	3.0	2.44	2.68	2.88	2.99	2.97	2.97	0.14	4.77
3	0.25	0.11	0.14	0.18	0.23	0.23	0.21	0.04	19.22
	1.0	0.53	0.65	0.81	0.94	0.97	0.96	0.14	15.06
	3.0	2.66	2.72	2.87	3.02	3.00	2.94	0.11	3.65

*= Three replicate measurements, SD =Standard Deviation, %RSD=%Relative Standard Deviation

4.1.1 Analytical method validation for flame retardants

4.1.1.1 Specificity examination:

The reagents or matrix used for flame retardants screening study may have conceivable interfering effect on chromatographic results. Reagent blank specimen, matrix blank specimen, and pure standard specimens were prepared and examined for interpreting any interfering effect. After examining method blank or matrix blank it was found that there was no any considerable amount of contamination present in the reagents or matrix. Equipment sensitivity check, specimen spike, laboratory quality check and calibration standard check were also examined. These test specimens shows results in the range 100 ± 5%. (table 3.3 can be explored). From the results is was found that the subjected method is very specific for screening flame retardants.

4.1.1.2 Working range & linearity examination:

For flame retardants test specimen's examination on equipment a six-point calibration curve was prepared and analyzed. Curve points 0.25 mg L^{-1}, 0.5 mg L^{-1}, 1.0 mg L^{-1}, 2.0 mg L^{-1}, 3.0 mg L^{-1} and 5.0 mg L^{-1} were prepared and examined on equipment. Regression value received for that curve was very efficient (0.996 to 0.997). It was linear. Its working range was 0.25 mg L^{-1} to 5.0 mg L^{-1}. These respective curve points were prepared three times separately. Each curve point was injected and analyzed three time on equipment (figure 4.1.2 and 4.1.3 can be explored for result interpretation)

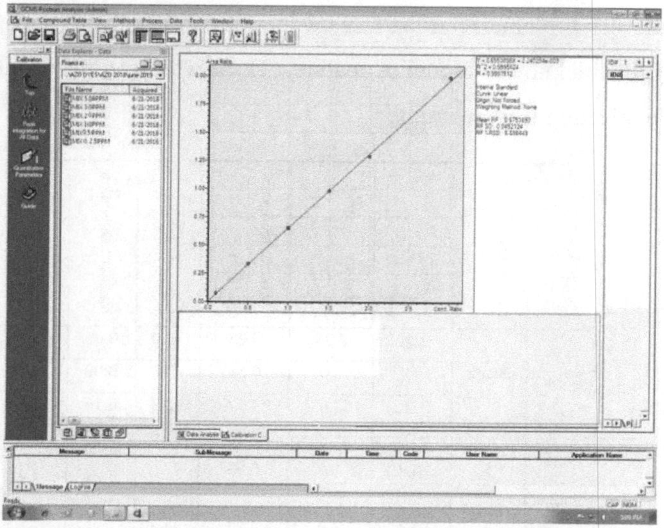

Figure 4.1.2 Calibration curve for flame retardants from GCMS instrument

curve data.pdf

Figure 4.1.3 Calibration curve data from GCMS instrument background

4.1.1.3 Accuracy:

For accuracy examination, four test specimens were analyzed. One of them was processed as blank, other three were spiked with 0.25 mg L^{-1}, 1.0 mg L^{-1} and 3.0 mg L^{-1} of pure standard concentrations. Each concentration level was injected and analyzed three time on equipment. These individually prepared replicates show recovery in the range 100± 10% (table 4.1.4 can be explored for result interpretation).

Table 4.1.4 Flame retardants recovery results (mg L^{-1}) when they were spiked with known amount of analyte. (These were average results of three-time injection from instrument)

Target Analyte	Test specimen	Quantity detected (mg L^{-1})	Quantity spiked (mg L^{-1})	Quantity recovered (mg L^{-1})*	Recovery (%)	Standard Deviation	%Relative Standard Deviation
Flame retardants	Blank	BDL	0.25	0.23	92.0	0.01	5.89
		BDL	1.0	0.89	89.0	0.08	8.23
		BDL	3.0	2.77	92.3	0.16	5.64
	1	BDL	0.25	0.22	84.0	0.02	9.03
		BDL	1.0	0.93	93.0	0.05	5.13
		BDL	3.0	2.83	94.3	0.12	4.12
	2	BDL	0.25	0.22	88.0	0.02	9.03
		BDL	1.0	0.94	94.0	0.04	4.37
		BDL	3.0	2.90	96.7	0.07	2.40
	3	BDL	0.25	0.24	96.0	0.01	2.89
		BDL	1.0	0.88	88.0	0.08	9.03
		BDL	3.0	2.91	97.0	0.06	2.15

BDL = Below Detection Limit,

*: Three replicate measurement

4.1.1.4 Precision: Three blank test specimen solutions were prepared. These solutions were spiked with pure standards. These solutions were divided into 10 parts each. In the next six days these 10 replicates from each concentration level were analyzed on equipment. Each replicate was analyzed on equipment three time. Relative standard deviation was examined for each concentration level. For the intra-day it was 0.05 to 0.12 % and for inter-days was 0.09 – 0.23 % (table 4.1.5 can be explored for result interpretation).

Table 4.1.5 Flame retardants result in mg L^{-1} for precision study. (These were average results for each replicate from 9-time examination)

Spiked concentration (mg L^{-1})	Intra-day precision			Inter-day precision		
	Recovery ± SD (mgL^{-1})[a]	% RSD (Precision)	Accuracy (%)	Recovery ± SD (mgL^{-1})[b]	% RSD (Precision)	Accuracy (%)
0.25	0.25±0.04	0.05	100.60	0.25±0.06	0.09	101
1.0	1.0±0.07	0.07	99.40	1.0±0.08	0.15	97
3.0	3.0±0.09	0.12	97.70	3.0±-0.13	0.23	96

[a] = Six replicate average results,

[b] = Nine replicate average results

4.1.1.5 LOD & LOQ:

To find limit of detection & limit of quantitation for the proposed analytical method blank and spike solutions were examined on GC-MS equipment. Spiked solutions were prepared on the basis of gradually falling concentration of analyte. These solutions were examined on equipment using developed analytical method. The lowest concentration of analyte that was detected successfully provide information about the LOD and the lowest concentration of analyte that was quantified successfully provide information about the LOQ. In alternative case signal-to-noise ratio was interpreted. For limit of detection calculation, the difference taken was 3:1 and for limit of quantitation calculation the difference taken was 10:1. In final calculations the limit of detection for analytes was 0.02 mg L^{-1} and limit of quantification was 0.2 mg L^{-1}.

4.1.1.6 Selectivity:

Selectivity for proposed analytical method was examined by preparing and analyzing specific test specimens. Specific test specimens were prepared by spiking different concentrations of analyte in the textile test specimens. In the presence of interfering matrix, the recovery for the flame retardants was found 100 ± 15% (above table 4.1.4 can be explored for result interpretation).

4.1.1.7 Robustness:

For the evaluation of the robustness of the proposed analytical method chromatographic parameters were varied intentionally. These parameters were changed in small amount. They were either increased or decreased. Temperature for column oven was changed intentionally from 50-340 °C to 45-335 °C and 55-345 °C. The carrier gas flow rate from 1.0 mL min^{-1} to 0.95 mL min^{-1} and 1.05 mL min^{-1}. It was found that method was efficiently stable to absorb these shocks. The results obtained after these variations shows that there was no any appreciable effect on the concentration of analyte. After evaluating analysis data, it was found that deviation was within 95 to 101% in the results (table 4.1.6 can be explored for result interpretation).

Table 4.1.6 Flame retardants result in mg L^{-1} for robustness study.

Test specimens	Quantity spiked (mg L^{-1})^{c}	Column oven temperature			Standard Deviation	%Relative Standard	Flow rate			Standard Deviation	%Relative Standard Deviation
		50-340 °C	45-335 °C	55-345 °C			1.0 mL/min	1.05 mL/min	0.95 mL/min		
A	0.25	0.24	0.25	0.26	0.01	3.27	0.24	0.26	0.23	0.01	5.27
B	1.0	1.01	1.03	1.04	0.02	1.79	0.97	0.96	1.02	0.03	2.79
C	3.0	2.96	2.95	2.97	0.02	0.73	3.03	3.04	3.02	0.02	0.57

^{c} = Three replicate measurements

4.1.1.8 Stability:

Analyte's stability was examined in the presence of other interference substances in the solution. Analyte's solution was prepared and analysis was performed in the subsequent three days period. Results were interpreted and % age of deviation was calculated for analytes. It was seen that deviation was < 4 %.

4.1.2 Test specimens screened for flame retardants:

For flame retardants, in the six-month period 232 test specimens were examined. After interpreting date of detectable test specimens for flame retardants it was found that flame retardants were detected in the range of 1.20 mg L^{-1} to 18.34 mg L^{-1} (table "4.1.7" identifies results for only detectable test specimens).

Table: 4.1.7 Results of detectable test specimens analyzed on GCMS instrument (mg L-1)

Test `specimen	Test specimen Description	Test specimen Color	Flame retardant detected	Results from GCMS (mg L-1)A	Results from GCMS (mg L-1)B	Mean (A+B/2)	Standard Deviation	%RSD
4	Child toy	Yellow	TCEP TCPP	6.63 4.94	6.35 4.43	6.49 4.69	0.20 0.36	3.05 7.70
5	Child toy	Blue	Penta-BB	6.68	6.83	6.76	0.11	1.57
6	Child toy	Purple, pink and black	Tetra-BDE TBECH BDBPP	9.26 9.94 4.06	9.03 9.40 4.62	9.15 9.67 4.34	0.16 0.38 0.40	1.78 3.95 9.12
7	Child toy	Black and blue color	TiBP BBMP	9.66 9.51	9.12 9.16	9.39 9.34	0.38 0.25	4.07 2.65
8	Child toy	Multi	TEP	3.65	3.32	3.49	0.23	6.70
9	Plastic button	Violet	TBP TEPA	4.66 5.66	4.44 5.17	4.55 5.42	0.16 0.35	3.42 6.40
10	Plastic button	Pink	TMP	8.61	8.01	8.31	0.42	5.11
11	Plastic button	Black and blue	PBEB TDCPP	18.34 3.58	18.92 3.84	18.63 3.71	0.41 0.18	2.20 4.96
12	Button from finished garment (Shirt)	Brown	TBEP TPP	8.43 3.58	7.93 3.86	8.18 3.72	0.35 0.20	4.32 5.32
13	Finished garment	Purple	TEHP	8.06	8.94	8.50	0.62	7.32
14	Finished garment (Hood)	Black, blue and pink	TEHP TOCP MDPP	8.26 6.93 4.06	8.45 6.21 4.55	8.36 6.57 4.31	0.13 0.51 0.35	1.61 7.75 8.05
15	Printed garment (Pant)	Red and black print	TMCP Penta-BDE	7.20 5.56	7.52 5.33	7.36 5.45	0.23 0.16	3.07 2.99
16	Leather gloves	Brown	TPCP TBB Hexa-BB	1.20 2.60 7.70	1.42 2.22 7.32	1.31 2.41 7.51	0.16 0.27 0.27	11.88 11.15 3.58
17	Leather jacket	Brown	DBHC-TCTD	2.53	2.26	2.40	0.19	7.97
18	Leather shoes	Black	TBBPA	3.82	3.17	3.50	0.46	13.15
19	Plastic part of leather shoes	Black	DBPP	16.32	16.20	16.26	0.08	0.52
20	Foam	Black	TDBPP	16.10	16.64	16.37	0.38	2.33
21	Foam	Yellow	V6 TBPP	12.30 15.70	12.11 15.34	12.21 15.52	0.13 0.25	1.10 1.64
22	Foam	Red	TXP	4.73	4.03	4.38	0.49	11.30
23	Foam	Yellow	TBPH	8.56	8.34	8.45	0.16	1.84
24	Foam	Black	Hexa-BDE	8.35	8.58	8.47	0.16	1.92

25	Foam	Yellow	Bis-TBBPA	8.67	8.73	8.70	0.04	0.49
26	Foam	Red	BTBPE	1.70	1.54	1.62	0.11	6.98
27	Foam	Yellow	HBCDD	8.56	8.66	8.61	0.07	0.82
28	Textile fabric	Black	Dechlorane-Plus	1.36 1.70	1.62 1.51	1.49 1.61	0.18 0.13	12.34 8.37
29	Textile fabric	Yellow	Hepta-BDE	4.73	4.36	4.55	0.26	5.76
30	Textile fabric	Red	Octa-BB	8.56	8.02	8.29	0.38	4.61
31	Textile fabric	Yellow	Octa-BDE	2.73	2.35	2.54	0.27	10.58
32	Textile fabric	Yellow	Deca-BB	8.56	8.42	8.49	0.10	1.17
33	Textile fabric	Red	Deca-BDE	2.79	2.93	2.86	0.10	3.46
34	Textile fabric	Yellow	DBDPE	3.56	3.63	3.60	0.05	1.38

RSD: Relative standard deviation

4.2 Phthalates contents:

It was found from the results that peak resolution was in better form in test specimens. All the subjected analytes show their peaks within 19.7 min. Their peaks were separated from each other and easily identifiable (figure "4.2.0" identifies results).

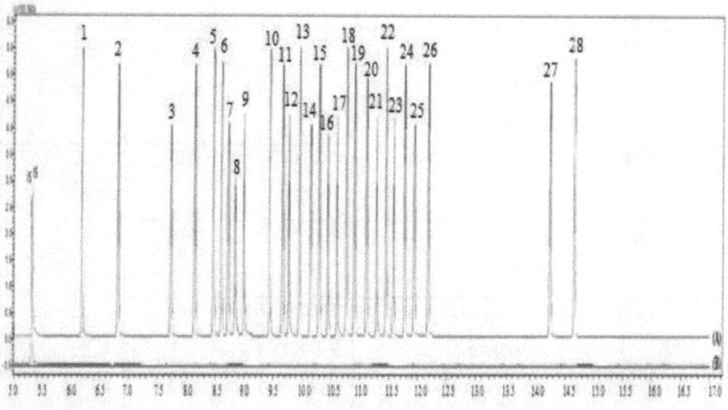

Figure 4.2.0 Reference retention times for phthalates from GC-MS chromatograms (a) matrix blank & (b) spiked test specimen (5.0 mg L^{-1}), Table 4.2.0 can be explored for 1-28 peaks identities, and section 3.7, 3.8 and 3.9.2 for analysis parameters.

Table 4.2.0 can be screened for mass fragments selected during time of analysis. Quantification of phthalates contents from test specimens were performed on the basis of internal standard method. For phthalates contents quantification anthracene d10 was used as internal standard. Structure and identity of each detected phthalates content was matched with the mass spectrum of standards mass spectrum (Figure "1.1 to 1.27" can be explored for result interpretation from annexure 2).

Table 4.2.0 phthalates contents and their mass fragment ions selected for analysis on GC-MS equipment.

Sr. No.	Short name	Full Name	CAS-No.	Target ion m/z on GC	Qualifier ions m/z on GC	Retention time (min)
1	DMP	Dimethyl phthalate	131-11-3	163	77, 194	6.22
2	DEP	Diethyl phthalate	84-66-2	149	177, 176	6.81
3	DPrP	Dipropyl phthalate	131-16-8	149	191	7.73
4	DIBP	Diiso butyl phthalate	84-69-5	149	223, 104	8.14
5	DMEP	Bis(2-methoxyethyl) phthalate	117-82-8	59	58, 149	8.45
6	DBP	Di-butyl phthalate	84-74-2	223	149, 167, 205	8.62
7	DIPP	Diisopentyl phthalate	605-50-5	149	237	8.70
8	nPIPP	n-Pentyl-isopentyl phthalate	776297-69-9	149	71, 237	8.82
9	DnPP	Di-n-pentyl phthalate	131-18-0	149	237	8.95
10	DIHP	Diisohexyl phthalate	71850-09-4	251	149	9.44
11	DnHP	Di-n-hexyl phthalate	84-75-3	149	251	9.68
12	DHP	Dihexyl phthalate	68515-50-4	251	149	9.75
13	BBP	Benzyl butyl phthalate	85-68-7	206	91.1, 149	9.98

14	DIHpP	Diisoheptyl phthalate	71888-89-6	99	149. 265	10.17
15	DEHP	Di-2-Ethyl Hexyl phthalate	117-81-7	279	149, 167	10.30
16	DnHpP	Di-n-heptyl phthalate	3648-21-3	150	149, 265	10.40
17	DPhP	Diphenyl phthalate	84-62-8	225	77, 104	10.55
18	DCHP	Dicyclohexyl phthalate	84-61-7	167	149, 249	10.75
19	DIOP	Diisooctyl phthalate	27554-26-3	113	149, 279	10.90
20	DDP	Dodecyl phthalate	84-77-5	149	307, 167	11.10
21	DBzP	Dibenzyl phthalate	523-31-9	107	149, 91	11.25
22	DnOP	Di-n-Octyl phthalate	117-84-0	279	149, 167, 261	11.42
23	DPHP	Di-(2-Propyl Heptyl) phthalate	53306-50-0	141	149, 167	11.60
24	DNP	Dinonyl phthalate	84-76-4	293	149, 167	11.80
26	DINP	Diisononyl phthalate	28553-12-0/ 68515-48-0	293	149, 167	11.92
25	DIDP	Diisodecyl phthalate	26761-40-0/ 68515-49-1	307	149, 167	12.20
27	DHNUP	Di (heptyl, nonyl, undecyl) phthalate	68515-42-4	149	150, 321	14.25
28	DUDP	Diundecyl phthalate	3648-20-2	149	150, 321	14.65

The test specimens that were collected from market were examined for target analytes and some test specimens as A, B and C from them were selected as blank matrix. In these test specimens' pure standards of phthalate contents (1.0 mg L^{-1}, 2.0 mg L^{-1} and 5.0 mg L^{-1}) were spiked for examining different parameters for the textile test specimens, child care items and toys. Various analysis parameters as adequate time for analysis, methods suitability for test specimens and digestion time duration etc. were optimized. Solvents as dichloromethane, tetrahydrofuran and chloroform, extraction methods as centrifuge, water both with shaker and ultrasonic bath, times as 20, 30, 40, 50, 60, and 70 minutes were used alternatively to reproduce subjected phthalate contents. Adequate solvent was found tetrahydrofuran, time 60 min and extraction method sonication. (figure "4.2.1" and table 4.2.1, 4.2.2, and 4.2.3 can be explored for result interpretation).

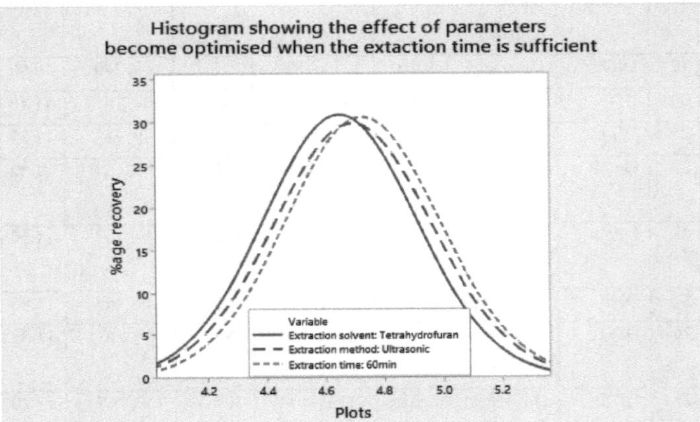

Figure 4.2.1 Results of textile test specimens spiked with phthalate contents concentration of 5.0 mg L^{-1} for examining the parameters as extraction solvent, method, time.

Table 4.2.1 Textile test specimens spiked with analyte concentration in mg L^{-1} for examining the extraction effect of different extraction solvents on phthalate contents.

Test specimen	Spike Level (mg L^{-1})	Extraction solvents		
		Chloroform	Dichloromethane	Tetrahydrofuran
A	1	0.72	0.64	0.88
	2	1.45	1.61	1.86
	5	4.18	4.33	4.73
B	1	0.35	0.58	0.68
	2	1.02	1.23	1.72
	5	3.48	4.18	4.48
C	1	0.33	0.41	0.86
	2	1.42	1.35	1.76
	5	2.98	3.56	4.57

Table 4.2.2 Textile test specimens spiked with analyte concentration in mg L^{-1} for examining the extraction effect of different extraction methods on phthalate contents.

Test specimen	Spike Level (mg L^{-1})	Extraction methods		
		Ultrasonic	Centrifuge	Water both with shaker
A	1	0.91	0.78	0.83
	2	1.93	1.65	1.75
	5	4.93	4.55	4.65
B	1	1.99	1.08	1.18
	2	2.05	2.12	2.21
	5	5.06	5.13	5.17
C	1	1.03	0.92	0.88
	2	2.06	2.14	2.19
	5	5.02	5.13	5.18

Table 4.2.3 Textile test specimens spiked with analyte concentration in mg L^{-1} for examining the extraction effect of different extraction time on phthalate contents.

Test specimen	Spike Level (mg L^{-1})	Extraction time					
		20 min	30 min	40 min	50 min	60 min	70 min
A	1	0.58	0.64	0.83	0.92	0.97	0.93
	2	1.61	1.68	1.88	1.89	1.99	1.94
	5	4.45	4.58	4.88	4.93	5.02	4.94
B	1	0.62	0.73	1.03	1.03	1.03	1.01
	2	1.52	1.63	1.70	1.77	1.95	1.94
	5	4.62	4.78	4.84	4.93	4.98	5.05
C	1	0.72	0.73	0.78	0.83	0.96	0.95
	2	1.65	1.88	1.95	1.99	2.03	2.09
	5	4.75	4.82	4.93	5.03	5.08	5.15

4.2.1 Analytical method validation for phthalate contents:

4.2.1.1 Specificity examination:

The reagents or matrix used for phthalate contents screening study may have conceivable interfering effect on chromatographic results. Reagent blank specimen, matrix blank specimen, and pure standard specimens were prepared and examined for interpreting any interfering effect. After examining method blank or matrix blank it was found that there was no any considerable amount of contamination present in the reagents or matrix. Equipment sensitivity check, specimen spike, laboratory quality check and calibration standard check were also examined. These test specimens shows results in the range 100 ± 5%. (table 3.3 can be explored). From the results is was found that the subjected method is very specific for screening phthalate contents.

4.2.1.2 Working range & linearity examination:

For phthalate contents test specimen's examination on equipment a six-point calibration curve was prepared and analyzed. Curve points 0.2 mg L^{-1}, 0.9 mg L^{-1}, 2.0 mg L^{-1}, 3.5 mg L^{-1}, 5.0 mg L^{-1} and 10.0 mg L^{-1} were prepared and examined on equipment. Regression value received for that curve was very efficient (0.996 to 0.997). It was linear. Its working range was 0.20 mg L^{-1} to 10.0 mg L^{-1}. These respective curve points were prepared three times separately. Each curve point was injected and analyzed three time on equipment (figure 4.2.2 and 4.2.3 can be explored for result interpretation)

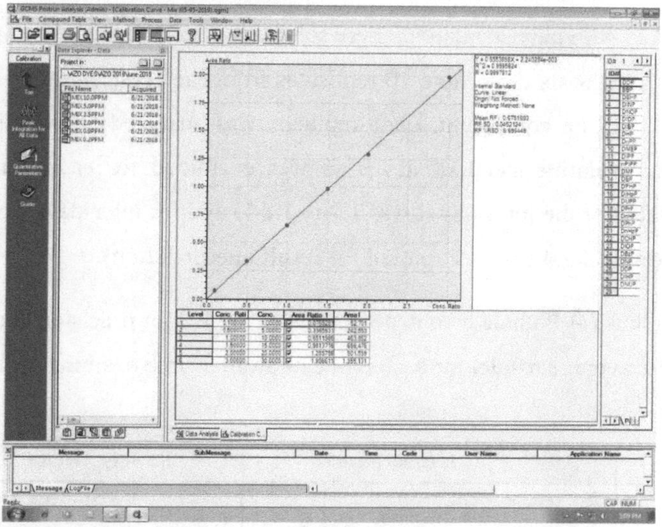

Figure 4.2.2 Calibration curve for phthalate contents from GCMS instrument

Mix PHS (05-05-2019).pdf

Figure 4.2.3 Calibration curve data from instrument background

4.2.1.3 Accuracy examination:

For accuracy examination, four test specimens were analyzed. One of them was processed as blank, other three were spiked with 1.0 mg L^{-1}, 2.0 mg L^{-1} and 5.0 mg L^{-1} of pure standard concentrations. Each concentration level was injected and analyzed three time on equipment. These individually prepared replicates show recovery in the range 100± 15% (table 4.2.1, 4.2.2 and 4.2.3 can be explored for result interpretation).

4.2.1.4 Precision:

Three blank test specimen solutions were prepared. These solutions were spiked with pure standards. These solutions were divided into 10 parts each. In the next six days these 10 replicates from each concentration level were analyzed on equipment. Each replicate was analyzed on equipment three time. Relative standard deviation was examined for each concentration level. For the intra-day it was 1.2 to 1.8 % and for inter-days was 2.2 – 2.8 % (table 4.2.4 can be explored for result interpretation).

Table 4.2.4 Phthalate contents results in mg L^{-1} for precision study. (These were average results for each replicate from 9-time examination)

Concentration spiked (mg L^{-1})	Inter-day precision			Intra-day precision		
	Recovery ± SD [c] (mg L^{-1})	% RSD (Precision)	Accuracy (%)	Recovery ± SD [b] (mg L^{-1})	% RSD (Precision)	Accuracy (%)
1.0	1.08±0.05	0.05	100.70	1.05±0.03	0.03	101.60
2.0	1.93±0.08	0.12	97.80	1.96±0.06	0.08	98.60
5.0	4.90±-0.14	0.21	95.90	4.94±-0.11	0.11	96.80

[b] = Six replicate average results

[c] = Nine replicate average results

4.2.1.5 LOD & LOQ examination:

To find limit of detection & limit of quantitation for the proposed analytical method blank and spike solutions were examined on GC-MS equipment. Spiked solutions of phthalate contents were prepared on the basis of gradually falling concentration of analytes. These solutions were examined on equipment using developed analytical method. The lowest concentration of analyte that was detected successfully provide information about the LOD and the lowest concentration of analyte that was quantified successfully provide information about the LOQ. In alternative case signal-to-noise ratio was interpreted. For limit of detection calculation, the difference taken was 3:1 and for limit of quantitation calculation the difference taken was 10:1. In final calculations the limit of detection for target analytes was in the range 0.03 – 0.05 mg L^{-1} and limit of quantification for target analytes was in the range 0.3 – 0.5 mg L^{-1}.

4.2.1.6 Selectivity examination:

Selectivity for proposed analytical method was examined by preparing and analyzing specific textile test specimens with commonly occurring interferences found in the toys, textile test specimens and child care items. Specific test specimens were prepared by spiking different concentrations of phthalate contents in the textile test specimens. In the presence of interfering matrix, the recovery for the phthalate contents was found in the range 100 ± 15% (above table 4.2.1, 4.2.2 and 4.2.3 can be explored for result interpretation).

4.2.1.7 Robustness:

For the evaluation of the robustness of the proposed analytical method chromatographic parameters were varied intentionally. These parameters were changed in small amount. They were either increased or decreased. Temperature for column oven was changed intentionally from 50-310 °C to

45-305 °C and 55-315 °C. The carrier gas flow rate from 3.0 mL/min to 2.95 mL/min and 3.05 mL/min. It was found that method was efficiently stable to absorb these shocks. The results obtained after these variations shows that there was no any appreciable effect on the concentration of analyte. After evaluating analysis data, it was found that deviation was within ±10 % in the results (table 4.2.5 can be explored for result interpretation).

Table 4.2.5 Phthalate contents results in mg L^{-1} for robustness study.

Test specimen	Quantity spiked (mg L^{-1})	Column oven temperature 50-310 °C	Column oven temperature 45-305 °C	Column oven temperature 55-315 °C	Flow rate, 3.0 mL/min	Flow rate, 3.05 mL/min	Flow rate, 2.95 mL/min
A	1.0	0.94	0.95	0.97	1.04	1.06	1.07
B	2.0	2.05	2.03	2.04	1.97	1.96	2.06
C	5.0	4.96	4.97	4.98	5.05	5.04	5.06

4.2.1.8 Stability:

Analyte's stability was examined in the presence of other interference substances in the solution. Analyte's solution was prepared and analysis was performed in the subsequent three days period. Results were interpreted and % age of deviation was calculated for analytes. It was seen that deviation was < 2 % in average.

4.2.1.9 Test specimens screened for phthalate contents:

For phthalate contents, in the three-month period 126 test specimens were examined. After interpreting date of detectable test specimens for phthalate contents it was found that Phthalate contents were detected in the range of 5.18 mg L^{-1} to 1798.14 mg L^{-1} (table "4.2.6" identifies results for only detectable test specimens).

Table 4.2.6. Phthalate contents (mg L^{-1}) results from detectable test specimens.

Sr. No.	Test specimen Color	Test specimen Description	Phthalate detected	GCMS Results mg L^{-1} [a]	GCMS Results duplicate test specimen mg L^{-1} [b]	Mean [a+b÷2]	Standard Deviation (SD)	%Relative Standard Deviation
D	Violet	Child toy	DONP DHP	40.66 5.66	39.89 5.50	40.28 5.58	0.54 0.11	1.35 2.03
E	Yellow	Child toy	DIDP DIOP	36.63 5.10	34.00 5.25	35.32 5.18	1.86 0.11	5.27 2.05
F	Blue	Child toy	DIBP	6.68	8.20	7.44	1.07	14.45
G	Pink	Child toy	BBP	8.52	8.55	8.54	0.02	0.25
H	Multi	Child toy	DEHP	13.65	13.16	13.41	0.35	2.58
I	Black and blue color	Child toy	DINP DIPP	29.66 9.51	27.76 8.34	28.71 8.93	1.34 0.83	4.68 9.27
J	Purple	Finished garment (Female Shirt)	DMP	37.03	39.02	38.02	1.41	3.70
K	Composite of three colors (black, blue and pink)	Finished garment (Hood)	DEP	60.64	63.50	62.07	2.02	3.26
L	Red and black print	Printed garment (Pant)	DPhP DPHP	72.20 45.56	78.56 43.37	75.38 44.47	4.50 1.55	5.97 3.48
M	Purple, pink and black	Child toy	DnPP DBP DINP	9.26 9.94 54.06	9.30 8.56 50.34	9.28 9.25 52.20	0.03 0.98 2.63	0.30 10.55 5.04
N	Black and blue	Plastic button	DMEP DNP	18.09 36.58	21.10 37.79	19.60 37.19	2.13 0.86	10.86 2.30
O	Brown	Button from finished garment (Shirt)	nPIPP	89.76	87.56	88.66	1.56	1.75
P	Black	Plastic part of leather shoes	DnHpP	1796.32	1799.96	1798.14	2.57	0.14
Q	Black	Child care item	DCHP	416.10	414.75	415.43	0.95	1.95

R	Yellow	Child care item	DBzP DPHP	132.36 175.70	136.06 174.03	134.21 174.87	2.62 1.18	1.95 0.68
S	Brown	Leather gloves	DIHpP DDP DnHP	14.20 23.60 47.70	11.80 26.60 45.4	19.60 21.29 37.19	2.13 1.92 0.86	10.86 9.00 2.30
T	Brown	Leather jacket	DPrP	238.53	227.74	233.14	7.63	3.27
U	Black	Leather shoes	DUPP	632.82	633.05	632.94	0.16	0.03
V	Yellow	Child care item	DHNUP	378.56	385.82	382.19	5.13	1.34
W	Red	Child care item	DIHP	244.73	238.42	241.58	4.46	1.85

4.3 Biocide contents:

It was found from the results that peak resolution was in better form in test specimens. All the subjected analytes show their peaks within 12.0 minutes. CIT elute at 10.65 min MIT at 2.75 min, OIT at 5.85 min and triclosan at 7.85 min (figure "4.3.0" identifies results for positively tested test specimens). Their peaks were separated from each other and easily identifiable.

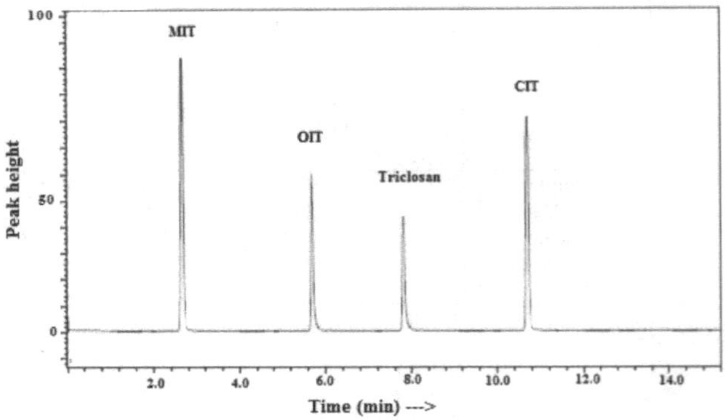

Figure 4.3.0. Shows LC peaks of biocides at 3 mg L^{-1} concentration in methanol

Table 4.3.0 Results of detectable test specimens for biocides analyzed on LC/DAD in mg kg^{-1}

Test specimens	Test specimen description	CIT (mg kg^{-1})	MIT (mg kg^{-1})	OIT (mg kg^{-1})	Triclosan (mg kg^{-1})
A	Black denim fabric	BDL	BDL	BDL	BDL
B	Blue denim fabric	BDL	BDL	BDL	BDL
C	Black fabric + foam	BDL	BDL	BDL	BDL
D	Black denim pant	BDL	BDL	6.53	2.93
E	Black denim pant	2.86	3.83	BDL	BDL
F	Black foam + fabric	8.95	4.38	10.99	12.44
G	Black denim fabric	10.83	BDL	3.92	BDL
H	Yellow fabric	25.35	BDL	BDL	2.99
I	Black denim pant	BDL	75.56	BDL	17.43
J	Brown fabric	BDL	10.45	BDL	BDL

The test specimens that were collected from market were examined for target analytes and some test specimens as A, B and C from them were selected as blank matrix. In these test specimens' pure standards of biocide contents (3.0 mg L^{-1}) were spiked for examining different parameters for the textile test specimens. Various analysis parameters as adequate time for analysis, methods suitability for test specimens and digestion time duration etc. were optimized. Solvents as water/methanol (1:1), acetonitrile/water (1:1), methanol, acetonitrile & water, extraction methods as centrifuge, water both with shaker and ultrasonic bath, times as 10, 20, 30, 40, 50 and 60 minutes were used alternatively to reproduce subjected biocide contents. Adequate solvent was found methanol, time 30 min and extraction method sonication. (figure 4.3.1 and table "4.3.1 to 4.3.3 can be explored for result interpretation).

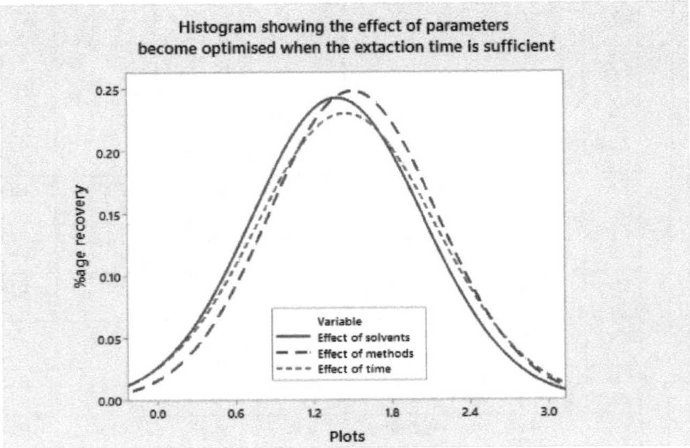

Figure 4.3.1 Results of textile test specimens spiked with biocide contents concentration of 3.0 mg L^{-1} for examining the parameters as extraction solvent, method & time.

Table 4.3.1 Textile test specimens spiked with analyte concentration 3.0 mg L^{-1} for examining the extraction effect of different extraction solvents on biocide contents.

Test specimen	Target analyte	Spiked amount 3.0 mg L^{-1}				
		Recovery (mg L^{-1}) in Methanol	Recovery (mg L^{-1}) in Acetonitrile	Recovery (mg L^{-1}) in Water	Recovery (mg L^{-1}) in Water/Methanol (1:1)	Recovery (mg L^{-1}) in Acetonitrile/Water (1:1)
A	MIT	2.89	2.25	2.45	2.82	2.84
	CIT	3.16	2.18	2.67	3.32	3.35
	OIT	3.23	2.67	2.26	2.68	2.64
	Triclosan	3.34	2.34	2.38	2.85	2.72
	MIT	2.97	2.62	2.67	2.99	3.19

B	CIT	3.02	3.34	3.18	3.13	3.26
	OIT	2.96	2.77	2.78	3.12	2.63
	Triclosan	2.96	2.78	2.88	2.85	2.91
C	MIT	3.05	2.21	8.87	2.77	3.03
	CIT	2.86	2.65	2.66	2.76	2.73
	OIT	3.15	2.83	2.76	2.78	3.17
	Triclosan	3.06	3.35	3.45	2.85	3.24

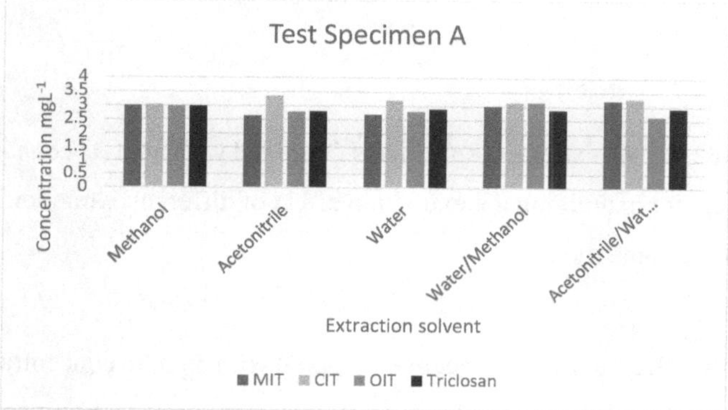

Figure 4.3.2 Textile test specimens A, spiked with analyte concentration 3.0 mg L^{-1} for examining the extraction effect of different extraction solvents on biocide contents.

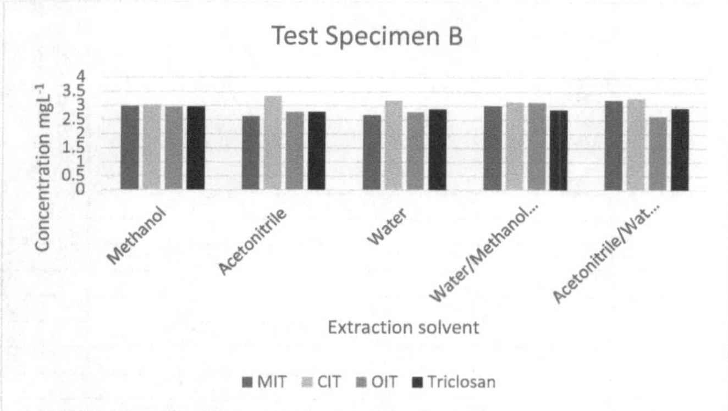

Figure 4.3.3 Textile test specimens B, spiked with analyte concentration 3.0 mg L^{-1} for examining the extraction effect of different extraction solvents on biocide contents.

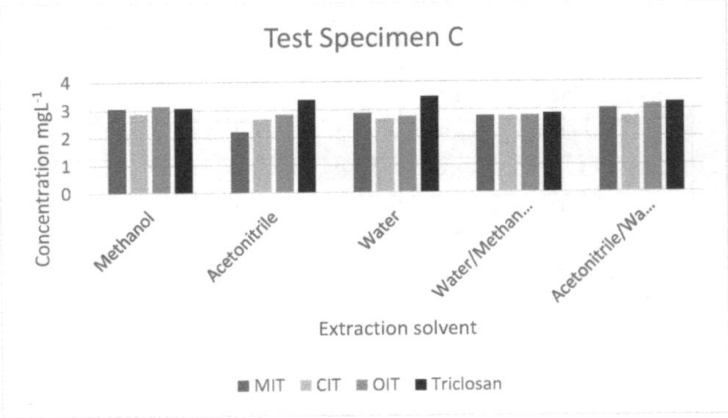

Figure 4.3.4 Textile test specimens C, spiked with analyte concentration 3.0 mg L^{-1} for examining the extraction effect of different extraction solvents on biocide contents.

Table 4.3.2 Textile test specimens spiked with analyte concentration 3.0 mg L^{-1} for examining the extraction effect of different extraction methods on biocide contents.

Test specimen	Target analyte	Spiked amount 3.0 mg L^{-1}		
		Recovery (mg L^{-1}) using Ultrasonic	Recovery (mg L^{-1}) using Centrifuge	Recovery (mg L^{-1}) using Water both with shaker
A	MIT	3.15	3.63	2.85
	CIT	3.25	2.51	2.67
	OIT	3.12	2.78	2.95
	Triclosan	3.17	2.88	3.12
B	MIT	3.14	3.11	2.83
	CIT	2.97	2.87	2.64
	OIT	3.01	2.89	2.71
	Triclosan	2.89	2.87	2.81
C	MIT	2.99	2.99	2.92
	CIT	2.88	2.87	2.73
	OIT	2.98	3.14	3.17
	Triclosan	2.77	2.56	2.15

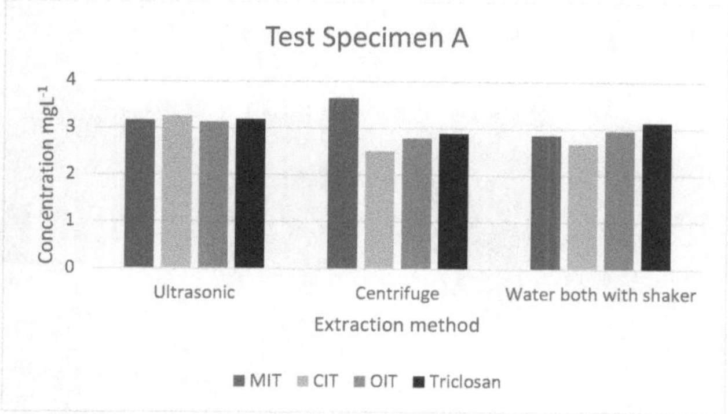

Figure 4.3.5 Textile test specimens A, spiked with analyte concentration 3.0 mg L^{-1} for examining the extraction effect of different extraction methods on biocide contents.

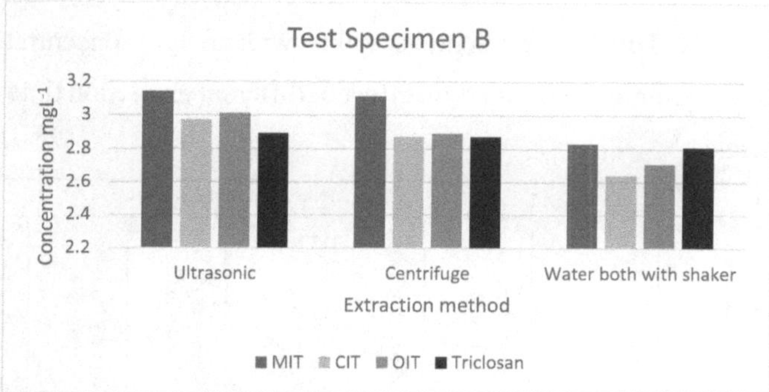

Figure 4.3.6 Textile test specimens B, spiked with analyte concentration 3.0 mg L^{-1} for examining the extraction effect of different extraction methods on biocide contents.

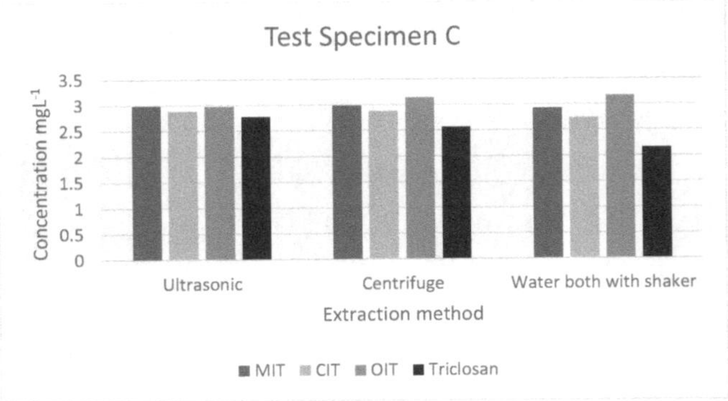

Figure 4.3.7 Textile test specimens C, spiked with analyte concentration 3.0 mg L^{-1} for examining the extraction effect of different extraction methods on biocide contents.

Table 4.3.3 Textile test specimens spiked with analyte concentration 3.0 mg L^{-1} for examining the extraction effect of different extraction time on biocide contents.

Test specimen	Target analyte	Spiked amount 3.0 mg L^{-1}					
		Recovery mg L^{-1} in 10 min	Recovery mg L^{-1} in 20 min	Recovery mg L^{-1} in 30 min	Recovery mg L^{-1} in 40 min	Recovery mg L^{-1} in 50 min	Recovery mg L^{-1} in 60 min
A	MIT	3.23	3.05	3.06	3.06	3.14	3.08
	CIT	2.61	2.88	2.97	2.96	3.05	3.06
	OIT	3.15	3.23	3.16	3.16	3.14	3.16
	Triclosan	3.22	3.16	3.15	3.25	3.22	3.17
B	MIT	2.74	2.87	2.98	2.97	2.99	2.93
	CIT	3.13	3.19	3.05	3.05	3.05	3.15
	OIT	2.75	2.83	2.99	2.95	2.97	2.98
	Triclosan	2.72	2.76	2.97	2.95	2.96	2.97
C	MIT	2.64	2.78	2.99	2.89	2.78	2.88
	CIT	3.21	3.35	3.14	3.16	3.28	3.22
	OIT	3.16	3.11	2.98	2.97	3.10	3.18
	Triclosan	3.14	3.16	3.06	3.11	3.13	3.10

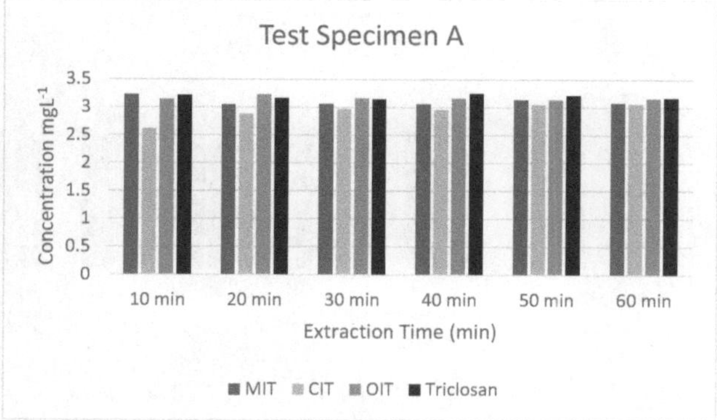

Figure 4.3.8 Textile test specimens A, spiked with analyte concentration 3.0 mg L^{-1} for examining the extraction effect of different extraction time on biocide contents.

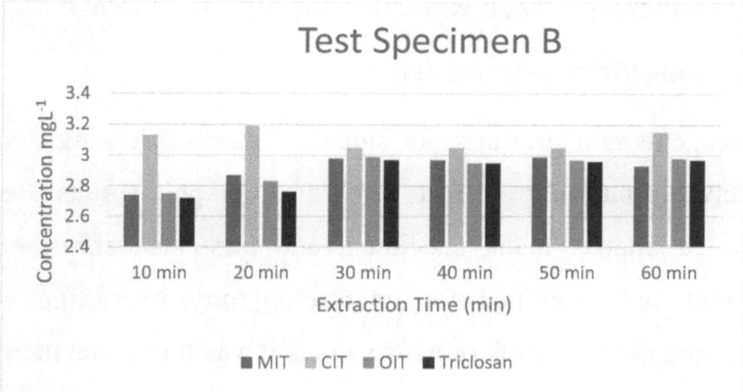

Figure 4.3.9 Textile test specimens B, spiked with analyte concentration 3.0 mg L^{-1} for examining the extraction effect of different extraction time on biocide contents.

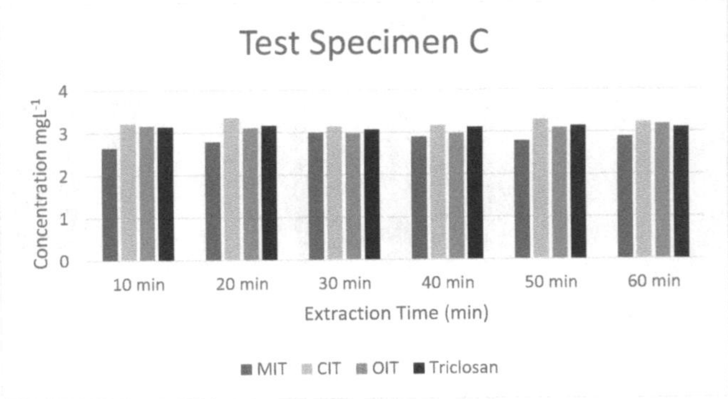

Figure 4.3.10 Textile test specimens C, spiked with analyte concentration 3.0 mg L^{-1} for examining the extraction effect of different extraction time on biocide contents.

4.3.1 Analytical method validation for biocide contents:

4.3.1.1 Specificity examination:

The reagents or matrix used for biocide contents screening study may have conceivable interfering effect on chromatographic results. Reagent blank specimen, matrix blank specimen, and pure standard specimens were prepared and examined for interpreting any interfering effect. After examining method blank or matrix blank it was found that there was no any considerable amount of contamination present in the reagents or matrix. Equipment sensitivity check, specimen spike, laboratory quality check and calibration standard check were also examined. These test specimens shows results in the range 100 ± 5%. (table 3.3 can be explored). From the results is was found that the subjected method is very specific for screening biocide contents.

4.3.1.2 Accuracy examination:

For accuracy examination, four test specimens were analyzed. One of them was processed as blank, other three were spiked with 1.0 mg L^{-1}, 3.0 mg L^{-1} and 5.0 mg L^{-1} of pure standard concentrations of biocide contents. Each concentration level was injected and analyzed three time on equipment. These individually prepared replicates show recovery in the range 100± 10% (figure 4.3.11 and table "4.3.4 can be explored for result interpretation).

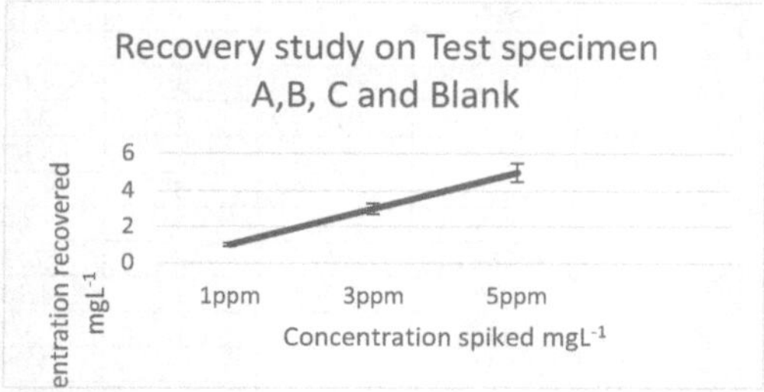

Figure 4.3.11 Textile test specimens A, B & C, spiked with analyte concentration in mg L^{-1} for recovery examining on biocide contents.

Table 4.3.4 Textile test specimens spiked with analyte concentration in mg L^{-1} for recovery examining on biocide contents.

Test specimen	Spike level	Target analyte	Quantity detected mg L^{-1}	Quantity Spiked mg L^{-1}	Quantity Found mg L^{-1}	Recovery %
Blank	Low	MIT	BDL	1	0.97	97.00
		CIT	BDL	1	0.96	96.00
		OIT	BDL	1	0.95	95.00
		Triclosan	BDL	1	1.08	108.00
	Medium	MIT	BDL	3	3.13	104.33
		CIT	BDL	3	3.06	102.00
		OIT	BDL	3	3.01	100.33
		Triclosan	BDL	3	2.98	99.33
	High	MIT	BDL	5	4.94	98.80
		CIT	BDL	5	4.97	99.40
		OIT	BDL	5	5.05	101.00

		Triclosan	BDL	5	5.13	102.60
A	Low	MIT	BDL	1	1.06	106.00
		CIT	BDL	1	1.05	105.00
		OIT	BDL	1	1.07	107.00
		Triclosan	BDL	1	0.98	98.00
	Medium	MIT	BDL	3	3.03	101.00
		CIT	BDL	3	3.06	102.00
		OIT	BDL	3	2.95	98.33
		Triclosan	BDL	3	2.96	98.67
	High	MIT	BDL	5	4.94	98.80
		CIT	BDL	5	4.98	99.60
		OIT	BDL	5	5.12	102.40
		Triclosan	BDL	5	5.06	101.20
B	Low	MIT	BDL	1	0.98	98.00
		CIT	BDL	1	0.97	97.00
		OIT	BDL	1	0.96	96.00
		Triclosan	BDL	1	1.09	109.00
	Medium	MIT	BDL	3	3.12	104.00
		CIT	BDL	3	3.08	102.66
		OIT	BDL	3	3.01	100.33
		Triclosan	BDL	3	2.97	99.00
	High	MIT	BDL	5	4.92	98.40
		CIT	BDL	5	4.97	99.40
		OIT	BDL	5	5.09	101.80
		Triclosan	BDL	5	5.13	102.60
C	Low	MIT	BDL	1	1.10	110.00
		CIT	BDL	1	1.07	107.00
		OIT	BDL	1	1.06	106.00
		Triclosan	BDL	1	0.95	95.00
	Medium	MIT	BDL	3	3.03	101.00
		CIT	BDL	3	3.08	102.66
		OIT	BDL	3	2.90	98.33
		Triclosan	BDL	3	2.96	96.66
	High	MIT	BDL	5	4.84	96.80
		CIT	BDL	5	4.93	98.60
		OIT	BDL	5	5.12	102.40
		Triclosan	BDL	5	5.21	102.20

BDL = Below Detection Limit

4.3.1.3 Precision examination: Three blank test specimen solutions were prepared. These solutions were spiked with pure standards. These solutions were divided into 10 parts each. In the next six days these 10 replicates from each concentration level were analyzed on equipment. Each replicate was analyzed on equipment three time. Relative standard deviation was examined for each concentration level. For the intra-day it was 1.8% to 2.8% and for inter-days was 2.4% – 3.9% (table 4.3.5 can be explored for result interpretation).

Table 4.3.5 Precision examination for biocide contents

Target analyte	Regression equation	Correlation coefficient	LOD (mg L^{-1})	LOQ (mg L^{-1})	Same day RSD (%) [a]	Six days RSD (%) [a]
CIT	Y=0.6460X-0.0308	0.9998	0.08	0.28	2.5	3.8
OIT	Y=0.4180X-0.0309	0.9997	0.12	0.38	2.8	3.9
MIT	Y=0.8360X-0.0598	0.9999	0.11	0.36	1.8	2.4
Triclosan	Y=0.3601X-0.0245	0.9995	0.06	0.21	1.9	2.9

Y = Peak area, X = Mean concentration (mg L^{-1}) Linear range = 1.0 - 5.0 mg L^{-1},

[a] =10 replicate measurements

4.3.1.4 LOD & LOQ examination:

To find limit of detection & limit of quantitation for the proposed analytical method blank and spike solutions were examined on LC-DAD equipment. Spiked solutions of biocide contents were prepared on the basis of gradually falling concentration of analytes. These solutions were examined on equipment using developed analytical method. The lowest concentration of analyte that was detected successfully provide information about the LOD and the lowest concentration of analyte that was quantified successfully provide information about the LOQ. In alternative case signal-to-noise ratio was interpreted. For limit of detection calculation, the difference taken was 3:1 and for limit of quantitation calculation the difference taken was 10:1. In final calculations the limit of detection for target analytes was in the range 0.06 – 0.12 mg L^{-1} and limit of quantification for target analytes was in the range 0.21 – 0.38 mg L^{-1} (above table 4.3.5 can be explored for result interpretation).

4.3.1.5 Working range & linearity examination:

For biocide contents test specimen's examination on equipment a five-point calibration curve was prepared and analyzed. Curve points 1.0 mg L^{-1}, 2.0 mg L^{-1}, 3.0 mg L^{-1}, 4.0 mg L^{-1} & 5.0 mg L^{-1} were prepared and examined on LC-DAD equipment. Regression value received for that curve was very efficient (0.9995 to 0.9999). It was linear. Its working range was 1.0 mg L^{-1} to 5.0 mg L^{-1}. These respective curve points were prepared three times separately. Each curve point was injected and analyzed three time on LC-DAD equipment (figure 4.3.12 and 4.3.13 can be explored for result interpretation)

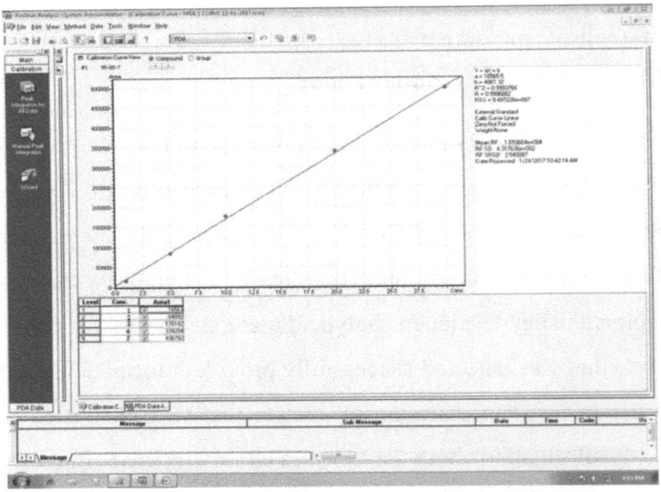

Figure 4.3.12 Calibration curve for biocide contents from LC-DAD instrument

BIO001.pdf

Figure 4.3.13 Calibration curve (data from GCMS instrument background)

4.3.1.6 Selectivity examination:

Selectivity for proposed analytical method was examined by preparing and analyzing specific textile test specimens with commonly occurring interferences found in the textile test specimens. Specific test specimens were prepared by spiking different concentrations of biocide contents in the textile test specimens. In the presence of interfering matrix, the recovery for the biocide contents was found in the range 100 ± 10% (above figure 4.3.12 and table "4.3.5" can be explored for result interpretation).

4.3.1.7 Robustness:

For the evaluation of the robustness of the proposed analytical method chromatographic parameters were varied intentionally. These parameters were changed in small amount. They were either increased or decreased. Temperature for column oven was changed intentionally ±3 °C. The carrier mobile phase flow rate from 1 mL/min to 0.9 mL and 1.1 mL/min. The concertation of mobile phase from 0.4 % Citric Acid: Methanol (50: 50) to 45:55. It was found that method was efficiently stable to absorb these shocks. The results obtained after these variations shows that there was no any appreciable effect on the concentration of analyte. After evaluating analysis data, it was found that deviation was within ±1 % in the results (table 4.3.6 and 4.3.7 can be explored for result interpretation).

Table 4.3.6 MIT results in mg L^{-1} for robustness study.

Analysis parameters	Retention Time, minutes	Assay, %	Number of theoretical plates	Tailing	Resolution
Column oven temperature 33	2.475	100.40	16234	1.34	9.65
Column oven temperature 27	2.511	100.38	18954	1.08	10.26
Flow rate, 1.1 mL min^{-1}	2.455	99.78	18225	1.06	9.85
Flow rate, 0.9 mL min^{-1}	2.535	100.33	17432	1.20	10.70
0.4% Citric Acid: Methanol (50: 50)	2.506	100.18	18340	0.96	10.36
0.4% Citric Acid: Methanol (55: 45)	2.485	99.42	18076	1.18	6.05
0.4% Citric Acid: Methanol (45: 55)	3.365	100.60	18897	1.09	12.28

Table 4.3.7 CIT results in mg L^{-1} for robustness study.

Analysis parameters	Retention Time, min	Assay, %	Number of theoretical plates	Tailing	Resolution
Column oven temperature 33	9.810	100.30	16554	1.24	9.56
Column oven temperature 27	9.925	100.25	18765	1.02	10.06
Flow rate, 1.1 mL min^{-1}	9.785	99.85	18034	1.04	9.63
Flow rate, 0.9 mL min^{-1}	9.945	100.54	17380	1.15	10.56
0.4% Citric Acid: Methanol (50: 50)	9.815	100.12	17834	0.94	10.12
0.4% Citric Acid: Methanol (55: 45)	9.735	99.63	18123	1.13	5.90
0.4% Citric Acid: Methanol (45: 55)	11.285	100.74	18435	1.06	12.22

4.3.1.8 Stability:

For biocide contents stability was examined in the presence of other interference substances in the solution. Biocide contents solutions were prepared and analysis was performed in the subsequent three days period. Results were interpreted and % age of deviation was calculated for analytes. It was seen that deviation was < 2 % in average.

4.3.2 Accelerated degradation on biocide contents:

For CIT, OIT, MIT and Triclosan rapid destruction studies was conducted. The HCl solution (0.1 N) was used for acid degradation of biocide contents at ambient temperature. 5 mL methanolic solution of biocide content was subjected for acidic degradation for 60.0 min time period in 1.5 mL HCl solution (0.1 N). Analyte's degradation was found to be 4-10% in acidic stress conditions. The NaOH solution (0.1 M) was used for alkaline degradation of biocide contents at ambient temperature. 5 mL methanolic solution of biocide content was subjected for alkaline degradation for 60.0 min time period in 1.5 mL NaOH solution (0.1 M). Analyte's degradation was found to be 3-12% in basic stress conditions. Heat decomposition was performed on the methanolic solution of biocide contents by exposing to high temperature. 5 mL methanolic solution of biocide content was subjected at high temperature of 80 °C for 72 hours. Analyte's degradation was found to be 6-14% in thermal stress conditions. For oxidative stress degradation 5 mL methanolic solution of biocide content was exposed to H_2O_2 (30% v/v) at ambient temperature for 60.0 min time period. Analyte's degradation was found to be 7-13% in oxidative stress conditions. 5 mL methanolic solution of biocide content was exposed to UV-light at 320 nm wavelength for 72 hours for photolytic tress degradation. Analyte's degradation was found to be 12-34% in photolytic stress conditions.

4.3.3 Test specimens screened for biocides:

For biocide contents, in the three-month period 135 test specimens were examined. After interpreting date of detectable test specimens for biocide contents it was found that biocide contents were detected in the range of 2.86 mg L^{-1} to 75.56 mg L^{-1} (table "4.3.8" identifies results for only detectable test specimens).

Table 4.3.8 Biocide contents (mg L^{-1}) results detected from multiple types of test specimens.

Test specimens	Test specimen description	MIT (mg kg^{-1})	CIT (mg kg^{-1})	OIT (mg kg^{-1})	Triclosan (mg kg^{-1})
A	Black fabric + foam	BDL	BDL	BDL	BDL
B	Black denim fabric	BDL	BDL	BDL	BDL
C	Blue denim fabric	BDL	BDL	BDL	BDL
D	Black denim pant	BDL	BDL	6.53	2.93
E	Black denim pant	3.83	2.86	BDL	BDL
F	Black foam + fabric	4.38	8.95	10.99	12.44
G	Black denim fabric	BDL	10.83	3.92	BDL
H	Yellow fabric	BDL	25.35	BDL	2.99
I	Black denim pant	75.56	BDL	BDL	17.43
J	Brown fabric	10.45	BDL	BDL	BDL

4.4 Ortho-Phenylphenol:

It was found from the results that peak resolution was in better form in detectable test specimens. The subjected analyte shows its peak within 17.0 min. OPP elute at 13.42 mins and internal standard Anthracene d-10 at 14.92 mins. Their peaks were separated from each other and easily identifiable (figure “4.4.0” identifies results).

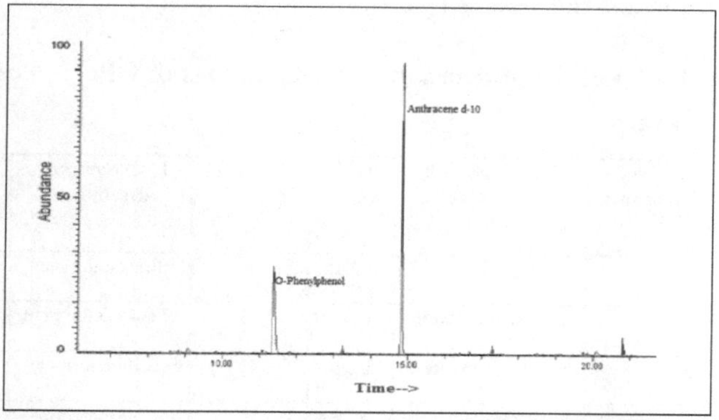

Figure 4.4.0 GC-MS chromatogram of O-Phenylphenol and internal standard (Anthracene d-10), section 3.7, 3.8 and 3.9.1 cab be explored for analysis parameters.

For OPP target ion was m/z 170 mass fragment and for Anthracene d-10 m/z 188 mass fragment (figure 4.4.1 identifies results for mass spectrum).

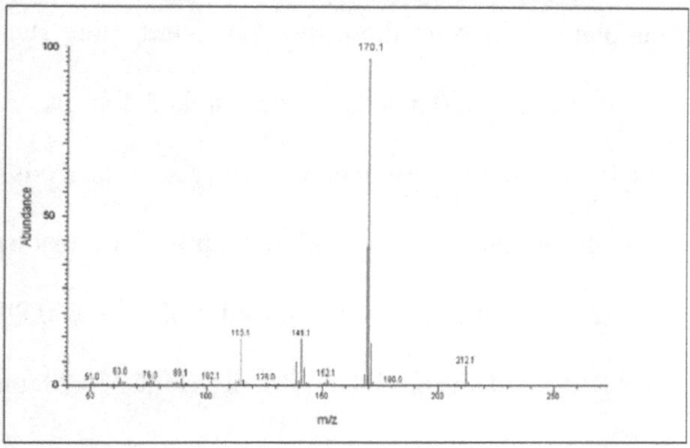

Figure 4.4.1 Mass spectrum of O-Phenylphenol

Quantification of OPP contents from the test specimens were performed on the basis of internal standard method. OPP contents in the range 1.15 mg L^{-1} to 35.52 mg L^{-1} were detected from the test specimens. Structure and identity of detected analyte was matched with the mass spectrum of standards specimen's mass spectrum (table 4.4.1 can be explored for detectable test specimen's result).

Table 4.4.1 Results of detectable test specimens of OPP contents analyzed on GC-MS

Test specimens	Test specimen's description	OPP mg L^{-1}	Test specimens	Test specimen's description	OPP mg L^{-1}
A	Black fabric	BDL	F	Blue denim pant	2.08
B	Blue denim fabric	BDL	G	Black denim pant	3.93
C	Black denim fabric	BDL	H	Brown leather	10.15
D	Black foam + fabric	4.28	I	Yellow fabric	3.77
E	Black leather	1.15	J	Black denim pant	35.52

The test specimens that were collected from market were examined for target analytes and some test specimens as A, B and C from them were selected as blank matrix. In these test specimens' pure standards of OPP contents (1.0 mg L^{-1}, 2.0 mg L^{-1}) were spiked for examining different parameters for the textile test specimens. Various analysis parameters as adequate time for analysis, methods suitability for test specimens and digestion time duration etc. were optimized. Solvents as 0.06 mg L^{-1} KOH, 0.08 mg L^{-1} KOH, 0.1 mg L^{-1} KOH, 0.1 mg L^{-1} KOH /Methanol (1:1) and 0.1 mg L^{-1} KOH /Acetonitrile (1:1), extraction methods as centrifuge, water both with shaker and ultrasonic bath, times as 20, 30, 40, 50, 60, 70 and 80 minutes were used alternatively to reproduce subjected OPP contents. Adequate solvent was found 0.1 mg L^{-1} KOH, time 60 min and extraction method sonication. (figure 4.4.2 and table 4.4.2, 4.4.3 and 4.4.4 can be explored for results interpretation).

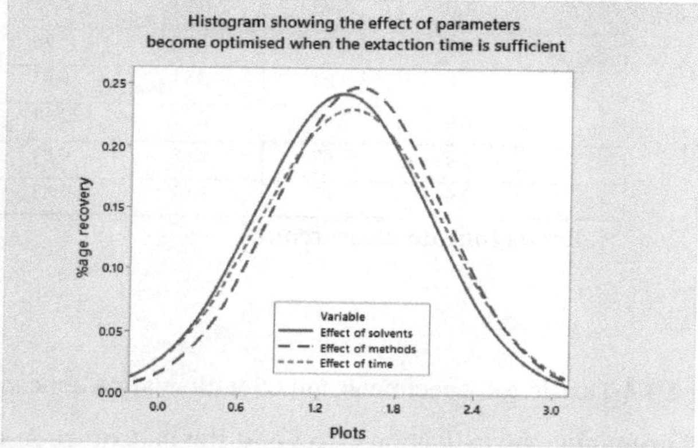

Figure 4.4.2 Results of textile test specimens spiked with OPP contents concentration of 2.0 mg L^{-1} for examining the parameters as extraction solvent, method, time.

Table 4.4.2 Textile test specimens spiked with analyte concentration in mg L^{-1} for examining the extraction effect of different extraction solvents on OPP contents.

Test specimen	Spike Level (mg L^{-1})^c	0.06 mg L^{-1} KOH	0.08 mg L^{-1} KOH	0.1 mg L^{-1} KOH	0.1 mg L^{-1} KOH /Methanol (1:1)	0.1 mg L^{-1} KOH /Acetonitrile (1:1)
A	1.0	0.92	0.93.	1.03	0.94	0.93
	2.0	1.85	1.88	1.99	1.91	1.90
B	1.0	0.91	0.90	1.06	0.95	0.94
	2.0	1.92	1.94	1.97	1.93	1.96
C	1.0	0.81	0.83	0.94	0.87	0.89
	2.0	1.74	1.69	1.86	1.81	1.79

^{c}: three replicate measurement

Table 4.4.3 Textile test specimens spiked with analyte concentration in mg L^{-1} for examining the extraction effect of different extraction methods on OPP contents.

Test specimen	Spike Level (mg L^{-1})^c	Ultrasonic	Centrifuge	Water both with shaker
A	1.0	1.05	0.93	0.85
	2.0	1.96	1.85	1.76
B	1.0	0.94	0.81	0.83
	2.0	2.07	2.17	2.14
C	1.0	1.07	0.88	0.84
	2.0	2.08	1.87	1.82

^{c}: three replicate measurement

Table 4.4.4 Textile test specimens spiked with analyte concentration in mg L^{-1} for examining the extraction effect of different extraction times on OPP contents.

Test specimen	Spike Level (mg L^{-1})^c	20 min	30 min	40 min	50 min	60 min	70 min	80 min
A	1.0	0.52	0.63	0.65	0.76	0.96	0.94	0.98
	2.0	1.41	1.51	1.58	1.87	1.94	1.97	1.96
B	1.0	0.64	0.74	0.87	0.98	1.07	1.09	1.03
	2.0	1.48	1.63	1.69	1.75	1.85	1.85	1.80
C	1.0	0.57	0.64	0.78	0.89	0.99	1.08	1.11
	2.0	1.41	1.44	1.65	1.74	1.86	1.88	1.82

^{c}: three replicate measurement

4.4.1 Analytical method validation for OPP contents:

4.4.1.1 Working range & linearity examination:

For OPP contents test specimen's examination on GC-MS equipment a six-point calibration curve was prepared and analyzed. Curve points 0.1 mg L^{-1}, 0.2 mg L^{-1}, 0.5 mg L^{-1}, 1.0 mg L^{-1}, 2.0 mg L^{-1} and 5.0 mg L^{-1} were prepared and examined on equipment. Regression value received for that curve was very efficient (0.9993 to 0.9995). It was linear. Its working range was 0.1 mg L^{-1} to 5.0 mg L^{-1}. These respective curve points were prepared three times separately. Each curve point was injected and analyzed three time on equipment (figure 4.4.3 can be explored for result interpretation)

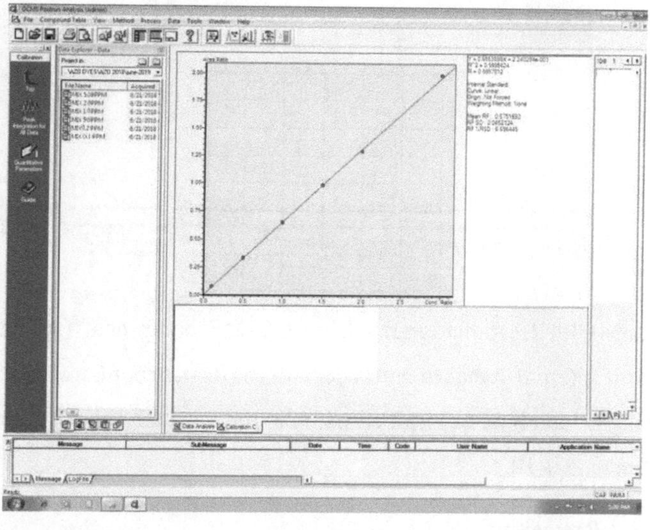

Figure 4.4.3 Calibration curve for OPP contents from GC-MS instrument

4.4.1.2 Specificity examination:

The reagents or matrix used for OPP contents screening study may have conceivable interfering effect on chromatographic results. Reagent blank specimen, matrix blank specimen, and pure standard specimens were prepared and examined for interpreting any interfering effect. After examining method blank or matrix blank it was found that there was no any considerable amount of contamination present in the reagents or matrix. Equipment sensitivity check, specimen spike, laboratory quality check and calibration standard check were also examined. These test specimens shows results in the range 100 ± 15%. (figure 4.4.4 & table 3.3 can be explored). From the results is was found that the subjected method is very specific for screening flame retardants.

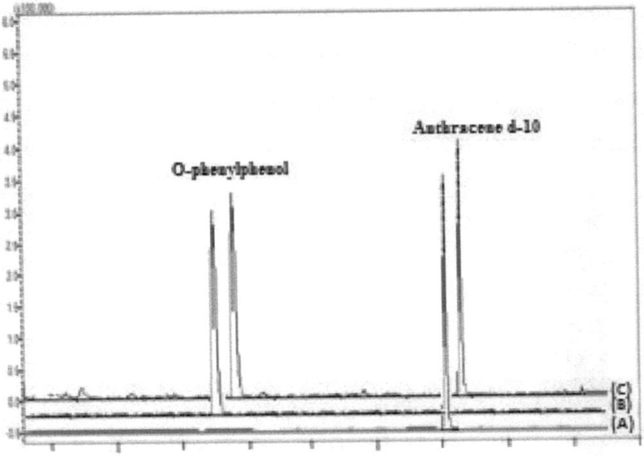

Figure 4.4.4 Chromatograms from GC-MS equipment (A) having single peak of internal standard anthracene d-10 (its matrix blank) (B) Spiked with 2.0 mg L^{-1} analyte, (C) Spiked with 2.0 mg L^{-1} analyte and internal standard anthracene d-10.

4.4.1.3 Precision:

Three blank test specimen solutions were prepared. These solutions were spiked with 3.0 mg L^{-1} pure standard of OPP contents. These solutions were divided into 10 parts each. In the next six days these 10 replicates from each concentration level were analyzed on GC-MS equipment. Each replicate was analyzed on equipment three times. Relative standard deviation was examined for each concentration level. For the intra-day it was 1.4% and for inter-days was 2.2% (table 4.4.5 can be explored for result interpretation).

Table 4.4.5 OPP contents result in mg L^{-1} for precision study.

Target analyte	Correlation coefficient	Regression equation	Same day RSD (%) [a]	Six day RSD (%) [a]	LOD (mg L^{-1})	LOQ (mg L^{-1})
OPP	0.9995	Y=0.7475X+2.1476	1.4	2.2	0.006	0.018

[a]=Ten replicate measurements, Linear range was 0.1 - 5.0 mg L^{-1}

4.4.1.4 Accuracy:

For accuracy examination, four test specimens were analyzed. One of them was processed as blank, other three were spiked with 1.0 mg L^{-1}, 3.0 mg L^{-1} and 5.0 mg L^{-1} of pure standard concentrations. Each concentration level was injected and analyzed three time on GC-MS equipment. These individually prepared replicates show recovery in the range 100± 15% (table 4.4.6 can be explored for result interpretation).

Table 4.4.6 Recovery results of OPP contents in mg L^{-1} for accuracy examination.

Target Analyte	Test specimen	Quantity detected (mg L^{-1})	Quantity spiked (mg L^{-1})	Quantity recovered (mg L^{-1})[c]	Recovery (%)
Ortho-phenylphenol (OPP)	**Blank**	BDL	0.1	0.098	98.0%
		BDL	0.2	0.188	94.0%
		BDL	1.0	0.97	97.0%
		BDL	5.0	4.92	98.4%
	A	BDL	0.1	0.095	95.0%
		BDL	0.2	0.190	95.0%

		BDL	1.0	0.89	89.0%
		BDL	5.0	4.86	97.2%
	B	BDL	0.1	0.092	92.0%
		BDL	0.2	0.191	95.5%
		BDL	1.0	0.93	93.0%
		BDL	5.0	4.81	96.2%
	C	BDL	0.1	0.088	88.0%
		BDL	0.2	0.173	86.5%
		BDL	1.0	0.91	91.0%
		BDL	5.0	4.73	94.6%

BDL = Below Detection Limit, [c]: three replicate measurement

4.4.1.5 LOD and LOQ:

To find limit of detection & limit of quantitation for the proposed analytical method blank and spike solutions were examined on GC-MS equipment. Spiked solutions were prepared on the basis of gradually falling concentration of analyte. These solutions were examined on equipment using developed analytical method. The lowest concentration of analyte that was detected successfully provide information about the LOD and the lowest concentration of analyte that was quantified successfully provide information about the LOQ. In alternative case signal-to-noise ratio was interpreted. For limit of detection calculation, the difference taken was 3:1 and for limit of quantitation calculation the difference taken was 10:1. In final calculations the limit of detection for analyte was 0.006 mg L^{-1} and limit of quantification was 0.018 mg L^{-1}.

4.4.1.6 Robustness:

For the evaluation of the robustness of the proposed analytical method chromatographic parameters were varied intentionally. These parameters were changed in small amount. They were either increased or decreased. Temperature for column oven was changed intentionally from 100-300 °C to 95-295 °C and 105-305 °C. The carrier gas flow rate from 3.0 mL min^{-1}

to 2.95 mL min^{-1} and 3.05 mL min^{-1}. It was found that method was efficiently stable to absorb these shocks. The results obtained after these variations shows that there was no any appreciable effect on the concentration of analyte. After evaluating analysis data, it was found that deviation was within 90 to 110% in the results (table 4.4.7 can be explored for result interpretation).

Table 4.4.7 Results of robustness study of OPP contents on GC-MS instrument.

Test specimen	Spike level (mg L^{-1}) ^{b}	Column oven temperature 100-300 °C	Column oven temperature 95-295 °C	Column oven temperature 105-305 °C	Flow rate, 3.0 mL min^{-1}	Flow rate, 3.05 mL min^{-1}	Flow rate, 2.95 mL min^{-1}
A	0.5	0.47	0.46	0.45	0.48	0.47	0.47
B	2.0	2.04	2.05	2.09	2.03	2.08	2.06
C	5.0	4.97	4.92	4.94	5.05	5.09	5.06

^{b}=Three replicate measurements

4.4.1.7 Stability examination:

Analyte's stability was examined in the presence of other interference substances in the solution. Analyte's solution was prepared and analysis was performed in the subsequent three days period. Results were interpreted and % age of deviation was calculated for analytes. It was seen that deviation was < 3 %.

4.4.1.8 Selectivity examination:

Selectivity for proposed analytical method was examined by preparing and analyzing specific test specimens. Specific test specimens were prepared by spiking different concentrations of OPP contents in the textile test specimens. In the presence of interfering matrix, the recovery for the OPP contents was found 100 ± 15% (above table 4.4.6 can be explored for result interpretation).

4.4.2 Textile samples screened for OPP contents:

For test specimens screened for OPP contents, 1.15 mg L^{-1} was the lowest value and 35.52 mg L^{-1} was the uppermost value. OPP contents 4.28 mg L^{-1} were detected in sample D, 1.15 mg L^{-1} in sample E, 2.08 mg L^{-1} in sample F, 3.93 mg L^{-1} in sample G, 10.15 mg L^{-1}, in sample H, 3.77 mg L^{-1} in sample I and 35.52 mg L^{-1} in sample J. In the five months period total 122 test specimens were analyzed; 7 test specimens show high detection (results were summarized in table 4.4.8).

Table 4.4.8 Results of detectable test specimens of OPP contents analyzed on GC-MS equipment.

Test specimens	Test specimen description	OPP mg L^{-1}	Test specimens	Test specimen description	OPP mg L^{-1}
A	Black fabric	BDL	F	Blue denim pant	2.08
B	Blue denim fabric	BDL	G	Black denim pant	3.93
C	Black denim fabric	BDL	H	Brown leather	10.15
D	Black foam + fabric	4.28	I	Yellow fabric	3.77
E	Black leather	1.15	J	Black denim pant	35.52

4.5 Dimethyl fumarate:

For the analyte content DMFu, in detectable specimen peak resolution was in better form. The analyte DMFu elute at 6.37 min time (results were summarized in figure 4.5.0).

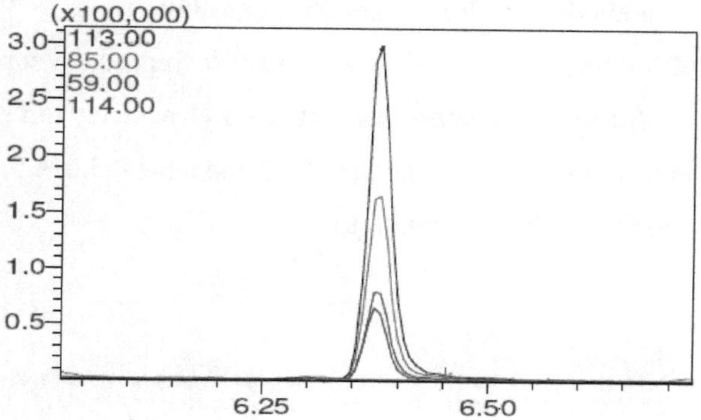

Figure 4.5.0 Chromatogram of DMFu from GCMS instrument, analytical method operated was presented at section 3.8 & 3.9.4.

Parent ion mass fragment m/z 113, daughter ions m/z 59 and m/z 85 were selected for DMFu as shown in mass spectrum of DMFu (Mass spectrum showed as figure 4.5.1).

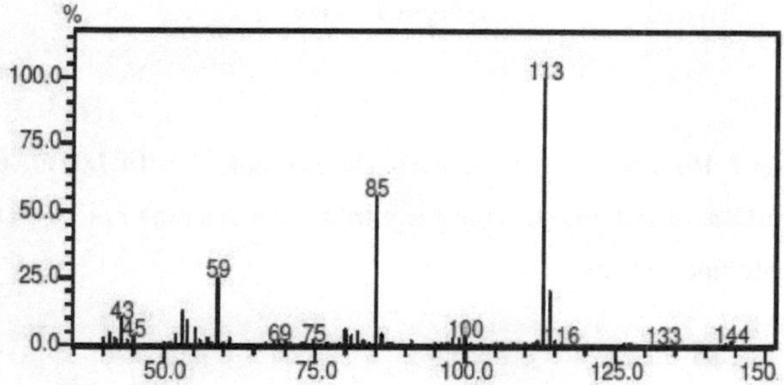

Figure 4.5.1 Mass spectrum of DMFu

The test specimens that were collected from market were examined for target analytes and some test specimens as A, B and C from them were selected as blank matrix. In these test specimens' pure standards of DMFu content (0.01 mg L^{-1}, 0.1 mg L^{-1} and 0.5 mg L^{-1}) were spiked for examining different parameters. Various analysis parameters as adequate time for analysis, methods suitability for test specimens and digestion time duration etc. were optimized. Solvents as ethyl acetate/methanol (1:1), ethyl acetate/acetonitrile (1:1), ethyl acetate and acetonitrile/methanol (1:1), extraction methods as reflux, rotary and sonication, times as 10, 20, 30, 40, 50 and 60 minutes were used alternatively to reproduce subjected flame retardants. Adequate solvent was found ethyl acetate, time 40 min and extraction method sonication. (figure 4.5.2 and table 4.5.0, 4.5.1 & 4.5.2 can be explored for result interpretation).

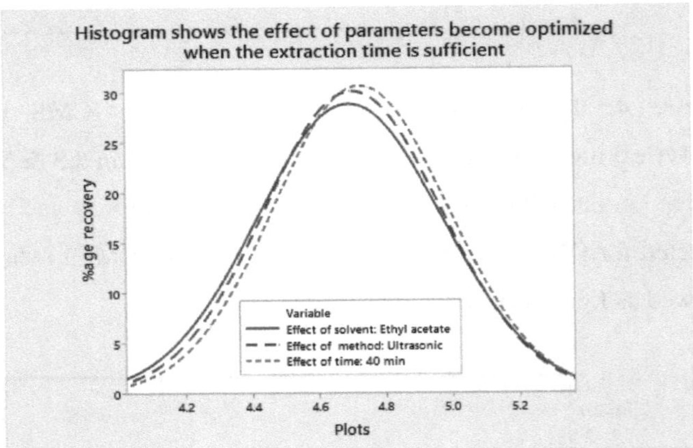

Figure 4.5.2 Results of textile test specimens spiked with DMFu contents concentration of 3.0 mg L^{-1} for examining the parameters as extraction solvent, method & time.

Table 4.5.0 Textile test specimens spiked with DMFu contents concentration in mg L^{-1} for examining the extraction effect of different extraction solvents.

Test specimen	Spike Level (mg/L)	Extraction solvents			
		Ethyl Acetate/Methanol (1:1)	Ethyl Acetate/Acetonitrile (1:1)	Ethyl Acetate	Acetonitrile/Methanol (1:1)
A	0.01	0.006	0.007	0.009	0.007
	0.1	0.072	0.084	0.093	0.068
	0.5	0.445	0.468	0.488	0.408
B	0.01	0.006	0.008	0.011	0.008
	0.1	0.078	0.083	0.095	0.073
	0.5	0.455	0.488	0.498	0.458
C	0.01	0.007	0.009	0.009	0.006
	0.1	0.092	0.093	0.103	0.088
	0.5	0.485	0.488	0.508	0.486

Table 4.5.1 Textile test specimens spiked with DMFu contents concentration in mg L^{-1} for examining the extraction effect of different extraction methods.

Test	Spike Level (mg L^{-1})	Extraction methods		
		Ultrasonic	Centrifuge	Water both with shaker
A	0.01	0.010	0.007	0.008
	0.1	0.091	0.078	0.083
	0.5	0.493	0.455	0.465
B	0.01	0.009	0.007	0.005
	0.1	0.096	0.077	0.082
	0.5	0.499	0.458	0.486
C	0.01	0.008	0.006	0.004
	0.1	0.105	0.090	0.092
	0.5	0.506	0.483	0.487

Table 4.5.2 Textile test specimens spiked with DMFu contents concentration in mg L^{-1} for examining the extraction effect of different extraction times.

Test specimen	Spike Level (mg/L)	Extraction time					
		10 min	20 min	30 min	40 min	50 min	60 min
A	0.01	0.003	0.004	0.006	0.009	0.009	0.008
	0.1	0.058	0.064	0.083	0.096	0.094	0.093
	0.5	0.465	0.468	0.488	0.498	0.499	0.494
B	0.01	0.004	0.006	0.008	0.0090	0.009	0.009
	0.1	0.048	0.063	0.085	0.095	0.088	0.095
	0.5	0.456	0.485	0.498	0.506	0.489	0.496
C	0.01	0.005	0.007	0.008	0.010	0.011	0.010
	0.1	0.062	0.073	0.103	0.103	0.095	0.108
	0.5	0.465	0.488	0.508	0.503	0.498	0.505

4.5.1 Analytical method validation for DMFu contents:

4.5.1.1 Working range and linearity examination:

For DMFu contents test specimen's examination on equipment a six-point calibration curve was prepared and analyzed. Curve points 0.01 mg L^{-1}, 0.025 mg L^{-1}, 0.05 mg L^{-1}, 0.10 mg L^{-1}, 0.25 mg L^{-1} and 0.5 mg L^{-1} were prepared and examined on equipment. Regression value received for that curve was very efficient (0.9995 to 0.9997). It was linear. Its working range was 0.01 mg L^{-1} to 0.5 mg L^{-1}. These respective curve points were prepared three times separately. Each curve point was injected and analyzed three time on equipment (figure 4.1.2 and 4.1.3 can be explored for result interpretation)

4.5.1.2 Specificity examination:

The reagents or matrix used for DMFu contents screening study may have conceivable interfering effect on chromatographic results. Reagent blank specimen, matrix blank specimen, and pure standard specimens were prepared and examined for interpreting any interfering effect. After examining method blank or matrix blank it was found that there was no any considerable amount of contamination present in the matrix or reagents. Equipment sensitivity check, specimen spike, laboratory quality check and calibration standard check were also examined. These test specimens shows results in the range 100 ± 15%. (table 3.3 can be explored). From the results is was found that the subjected method is very specific for screening DMFu contents.

4.5.1.3 Accuracy:

For accuracy examination, four test specimens were analyzed. One of them was processed as blank, other three were spiked with 0.1 mg L^{-1}, 0.3 mg L^{-1} and 0.5 mg L^{-1} of pure standard concentrations. Each concentration level was injected and analyzed three time on equipment. These individually prepared replicates show recovery in the range 91 % to 94 % (table 4.5.3 can be explored for result interpretation).

Table 4.5.3 DMFu contents recovery results (mg L^{-1})

Target analyte	Test specimens	Quantity detected (mg L^{-1})	Quantity spiked (mg L^{-1})	Quantity recovered (mg L^{-1})^c	Recovery (%)
Dimethyl fumarate (DMFu)	**Blank**	BDL	0.1	0.098	98.0%
		BDL	0.3	0.288	96.0%
		BDL	0.5	0.497	99.4%
	A	BDL	0.1	0.092	92.0%
		BDL	0.3	0.291	97.0%
		BDL	0.5	0.490	98.0%
	B	BDL	0.1	0.089	89.0%
		BDL	0.3	0.286	95.3%
		BDL	0.5	0.492	98.4%
	C	BDL	0.1	0.091	91.0%
		BDL	0.3	0.293	97.7%
		BDL	0.5	0.481	96.2%

BDL = Below Detection Limit, ^{c}: three replicate measurement

4.5.1.4 Precision examination:

For precision examination three blank test specimen solutions were prepared. These solutions were spiked with pure standards. These solutions were divided into 10 parts each. In the next six days these 10 replicates from each concentration level were analyzed on equipment. Each replicate was analyzed on equipment three time. Relative standard deviation was examined for each concentration level. For the intra-day it was 1.6 % and for inter-days was 2.5 %.

4.5.1.5 LOD and LOQ examination:

To find limit of detection & limit of quantitation for the proposed analytical method blank and spike solutions were examined on GC-MS equipment. Spiked solutions were prepared on the basis of gradually falling concentration of analyte. These solutions were examined on equipment using developed analytical method. The lowest concentration of analyte that was detected successfully provide information about the LOD and the lowest concentration of analyte that was quantified successfully provide information about the LOQ. In alternative case signal-to-noise ratio was interpreted. For limit of detection calculation, the difference taken was 3:1 and for limit of quantitation calculation the difference taken was 10:1. In final calculations the limit of detection for analytes was 5×10^{-4} mg L^{-1} and limit of quantification was 5×10^{-3} mg L^{-1}.

4.5.1.6 Selectivity: Selectivity for proposed analytical method was examined by preparing and analyzing specific test specimens. Specific test specimens were prepared by spiking different concentrations of analyte in the textile test specimens. In the presence of interfering matrix, the recovery for the DMFu contents was found 100 ± 15% (above table 4.5.3 can be explored for result interpretation).

4.5.1.7 Stability: Analyte's stability was examined in the presence of other interference substances in the solution. Analyte's solution was prepared and analysis was performed in the subsequent three days period. Results were interpreted and % age of deviation was calculated for analytes. It was seen that deviation was < 3 %.

4.5.1.8 Robustness: For the evaluation of the robustness of the proposed analytical method chromatographic parameters were varied intentionally. These parameters were changed in small amount. They were either increased or decreased. Temperature for column oven was changed intentionally from 50-250 °C to 45-245 °C and 55-255 °C. The carrier gas flow rate from 2.0 mL min^{-1} to 1.95 mL min^{-1} and 2.05 mL min^{-1}. It was found that method was efficiently stable to absorb these shocks. The results obtained after these variations shows that there was no any appreciable effect on the concentration of analyte. After evaluating analysis data, it was found that deviation was within 85 to 115% in the results (table 4.5.4 can be explored for result interpretation).

Table 4.5.4 Results of robustness study of DMFu form GCMS instrument.

Test specimen	Spike level (mg L^{-1})[a]	Column oven temperature 100-250 °C	Column oven temperature 95-245 °C	Column oven temperature 105-255 °C	Flow rate, 2.0 mL/min	Flow rate, 2.05 mL/min	Flow rate, 1.95 mL/min
A	0.1	0.096	0.081	0.085	0.088	0.079	0.078
B	0.2	0.203	0.205	0.207	0.201	0.206	0.205
C	0.5	0.495	0.492	0.494	0.503	0.508	0.506

[a]=Three replicate measurements

4.5.2 Textile samples screened for DMFu contents:

1.34 mg L^{-1} was the lowest value for the analyte DMFu and 5.12 mg L^{-1} was the uppermost detected value. Sample D contain 2.23 mg L^{-1} DMFu content, sample E contain 5.09 mg L^{-1}, sample F contain 2.11 mg L^{-1}, sample G contain 1.99 mg L-1, sample H contain 1.34 mg L^{-1}, sample I contain 4.65 mg L^{-1} and sample J contain highest value 5.12 mg L^{-1} of DMFu contents. In the four months period total 65 test specimens were analyzed; 7 test specimens show high detection (results were summarized in table 4.5.5).

Table 4.5.5 Results of detectable test specimens of DMFu analyzed on GCMS

Test specimens	Test specimen description	DMFu mg L^{-1}	Test specimens	Test specimen description	DMFu mg L^{-1}
A	PU Leather	BDL	F	Blue denim pant	2.11
B	Rubber	BDL	G	Black denim pant	1.96
C	Silica gel	BDL	H	Brown Leather	1.34
D	Rubber	2.23	I	Yellow fabric	4.65
E	PU Leather	5.09	J	Silica gel	5.12

4.6 Azo dyes (Aromatic amines):

Method was applied on 867 test specimens of 600 real articles. Textile commercial products, as finished garments, powder azo dyes, liquid azo dyes, printed and non-printed fabrics were collected for experimental examination from the textile and leather manufacturing industries and retailer shops in the Lahore city of Pakistan. Amines released from detected test specimens were in the range of 5 ppm to 1800 ppm. Finally, 22/867 test specimens contain amines above EU regulations i.e. $\geq$ 30 ppm and 168 test specimens contain amines at low level i.e. ≤ 30 ppm. Results of amines released from test specimens were summarized in table 4.6.5

Analyzed test specimens were of different compositions, they contain composite or single color but not more than three colors were mixed, one test specimen may contain multiple or a single amine. For results verification, each detected test specimen was run in duplicate and for each duplicate test specimen the relative standard deviation was ≤ 30%. Detected amine was also spiked with pure standard amine and test specimen was analyzed on GC/MS and UFLC. Spiking with pure amine clearly separate the false peak if present.

Test specimens contain interferences which interfere with the amines or internal standard peak. Interferences increase or decrease the area of internal standards (Naphthalene d-8 and

Anthracene d-10) or amines and produce the effect like they drop down the spectrum similarity of internal standard or amines.

Polyester test specimens with high-risk colors as red, black, blue or orange contain the interferences and these interferences interfere with internal standards Anthracene d-10 and Naphthalene d-8 as they reduce the peak area or drop down the internal standards spectrum similarity. To overcome such a problem, we change the organic solvent used for amine extraction as instead of Ethyl Acetate (EA) we use Tertiary Butyl Methyl Ether (TBME) or any other solvent as acetonitrile. Leather and Polyurethane (PU) test specimens contain some proteinaceous material or adhesives which interfere with internal standards or amines peaks. Some amines are interconvertible, sensitive exposure to hot environment, and any delay decrease the result of one amine and increase the result of other one, as any delay or exposure to hot environment decrease the concentration of 92-87-5 and increase the concentration of 92-67-1, test specimen number 6, 19, 24 and 25 in table 6, similarly 95-80-7 degrade and convert to 99-55-8 and 106-47-8, test specimen number 9, 10, 15 and 16 in table 6. We analyzed the test specimens immediately after processing to overcome such problem and confusion in results. Test specimens that contain very high concentration of amines and exceeds our calibration curve maximum value that is 30 ppm, we diluted such test specimens to bring the value in the range of calibration curve.

4.6.1 Analytical method validation for azo dyes (Aromatic amines):

4.6.1.1 Specificity examination: The reagents or matrix used for aromatic amines screening study may have conceivable interfering effect on chromatographic results. Reagent blank specimen, matrix blank specimen, and pure standard specimens were prepared and examined for interpreting any interfering effect. After examining method blank or matrix blank it was found that there was no any considerable amount of contamination added from these test specimens. Equipment sensitivity check, specimen spike, laboratory quality check and calibration standard check were also examined. These test specimens shows results in the range 100 ± 15%. (table 4.6.0 can be explored). From the result it was found that the subjected method is very specific for screening aromatic amines.

4.6.1.2 Working range & linearity examination:

For aromatic amine's test specimen examination on equipment a six-point calibration curve was prepared and analyzed. Curve points 1.0 mg L^{-1}, 5.0 mg L^{-1}, 10.0 mg L^{-1}, 15.0 mg L^{-1}, 20.0 mg L^{-1} and 30.0 mg L^{-1} were prepared and examined on GC-MS & UFLC equipment. Regression value received for that curve was very efficient (0.994 to 0.996). It was linear. Its working range was 1.0 mg L^{-1} to 30.0 mg L^{-1}. These respective curve points were prepared three times separately. Each curve point was injected and analyzed three time on equipment (table 4.6.0 can be explored for result interpretation)

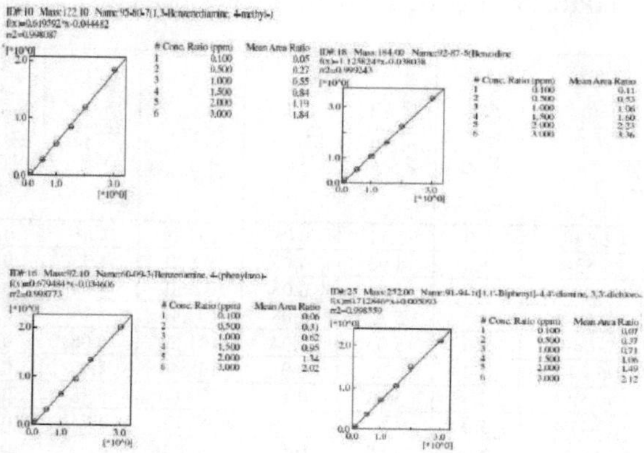

Fig. 4.6.0 Calibration curves of some mostly detected aromatic amines

(Data from instrument)

4.6.1.3 Accuracy examination: For accuracy examination, four test specimens were analyzed. One of them was processed as blank, other three were spiked with 5.0 mg L^{-1}, 15.0 mg L^{-1} and 30.0 mg L^{-1} of pure standard concentrations. Each concentration level was injected and analyzed three time on equipment. These individually prepared replicates show recovery in the range 100±5 % (table 4.6.0 can be explored for result interpretation).

Table 4.6.0 Recoveries of aromatic amines from textile test specimens

Sr. No.	Target amine	Quantity spiked (mg L^{-1})	GCMS Results (mg L^{-1})	SD	Recovery %	UFLC Results (mg L^{-1})	SD	Recovery %
1	62-53-3	5.0	5.11	0.08	102.20	5.08	0.06	101.60
2	95-53-4	5.0	4.96	0.03	99.20	4.93	0.05	98.60
3	95-68-1	5.0	4.88	0.08	97.60	4.84	0.11	96.80
4	87-62-7	5.0	5.12	0.08	102.40	5.09	0.06	101.80
5	90-04-0	5.0	5.06	0.04	101.20	5.08	0.06	101.60
6	106-47-8	5.0	5.02	0.01	100.40	4.95	0.04	99.00
7	106-50-3	5.0	5.23	0.16	104.60	5.18	0.13	103.60
8	120-71-8	5.0	4.87	0.09	97.40	4.85	0.11	97.00
9	137-17-7	5.0	4.84	0.11	96.80	4.91	0.06	98.20
10	95-69-2	5.0	5.02	0.01	100.40	5.11	0.08	102.20
11	95-80-7	5.0	5.13	0.09	102.60	5.04	0.03	100.80
12	615-05-4	5.0	5.18	0.13	103.60	5.21	0.15	104.20
13	91-59-8	5.0	4.97	0.02	99.40	5.13	0.09	102.60
14	99-55-8	5.0	4.89	0.08	97.80	4.87	0.09	97.40
15	92-67-1	5.0	4.93	0.05	98.60	5.07	0.05	101.40
16	60-09-3	5.0	5.01	0.01	100.20	5.09	0.06	101.80
17	101-80-4	5.0	5.13	0.09	102.60	5.07	0.05	101.40
18	92-87-5	5.0	5.10	0.07	102.00	5.03	0.02	100.60
19	101-77-9	5.0	5.15	0.11	103.00	5.11	0.08	102.20
20	97-56-3	5.0	5.07	0.05	101.40	5.13	0.09	102.60
21	838-88-0	5.0	5.06	0.04	101.20	4.95	0.04	99.00
22	119-93-7	5.0	5.04	0.03	100.80	4.96	0.03	99.20
23	139-65-1	5.0	5.05	0.04	101.00	4.94	0.04	98.80
24	101-14-4	5.0	5.14	0.10	102.8	5.18	0.13	103.60
25	119-90-4	5.0	5.03	0.02	100.6	5.08	0.06	101.60
26	91-94-1	5.0	5.07	0.05	101.4	5.13	0.09	102.60

4.6.1.4 Precision examination: Three blank test specimen solutions were prepared. These solutions were spiked with pure standards of aromatic amines. These solutions were divided into 10 parts each. In the next six days these 10 replicates from each concentration level were analyzed on GC-MS & UFLC equipment. Each replicate was analyzed on equipment three time. Relative standard deviation was examined for each concentration level. For the intra-day it was 2.5 to 3.0 % and for inter-days was 3.0 – 4.0 %.

4.6.1.5 LOD (limit of detection): 0.1 ppm was the LOD for this method.

4.6.1.6 LOQ (limit of quantification): 0.3 ppm was the LOQ value for this method.

4.6.1.7 Selectivity examination:

Selectivity for proposed analytical method was examined by preparing and analyzing specific test specimens. Specific test specimens were prepared by spiking different concentrations of aromatic amines in the textile test specimens. In the presence of interfering matrix, the recovery for the aromatic amines was found 100 ± 15%.

4.6.1.8 Robustness examination:

For the evaluation of the robustness of the proposed analytical method chromatographic parameters were varied intentionally. These parameters were changed in small amount. They were either increased or decreased. Temperature for column oven was changed intentionally from 36 °C to 34 °C & 38 °C on UFLC equipment and 100-280 °C to 95-175 °C and 105-285 °C on GC-MS equipment. The flow rate from 1.0 mL min^{-1} to 0.95 mL min^{-1} & 1.05 mL min^{-1} on GC-MS equipment and 0.70 mL/min to 0.65 mL/min & 0.75 mL/min on UFLC equipment. It was found that method was efficiently stable to absorb these shocks. The results obtained after these variations shows that there was no any appreciable effect on the concentration of analyte. After evaluating analysis data, it was found that deviation was within 95% to 101% in the results (table 4.6.1 & 4.6.2 can be explored for result interpretation).

Table 4.6.1 Results of robustness study of aromatic amines for UFLC instrument.

Sr. No.	Target amine	Column oven temperature 36 °C	Column oven temperature 34 °C	Column oven temperature 38 °C	Flow rate, 0.7 mL/min	Flow rate, 0.65 mL/min	Flow rate, 0.75 mL/min
1	62-53-3	5.06	5.03	5.04	4.97	4.96	5.03
2	95-53-4	5.04	5.06	5.03	5.07	5.04	4.94
3	95-68-1	4.97	4.98	4.96	4.95	4.93	4.94
4	87-62-7	5.02	5.08	5.06	5.03	5.02	4.97
5	90-04-0	4.99	4.95	5.06	5.05	5.07	5.06
6	106-47-8	5.04	5.03	5.04	5.04	5.08	5.07
7	106-50-3	5.05	5.06	5.07	5.06	5.09	5.07
8	120-71-8	4.98	5.05	4.99	5.06	5.08	5.05
9	137-17-7	4.97	4.94	4.95	5.04	5.06	5.07
10	95-69-2	5.04	5.05	5.03	4.97	4.96	5.06
11	95-80-7	4.97	4.96	4.97	5.05	5.04	5.06
12	615-05-4	4.98	4.97	4.96	5.07	5.08	5.09
13	91-59-8	5.03	5.05	5.02	4.96	4.95	4.95
14	99-55-8	5.06	5.05	5.04	4.98	4.97	4.95
15	92-67-1	5.06	5.08	5.07	4.96	4.94	4.95
16	60-09-3	4.99	5.04	5.05	5.06	5.07	5.08
17	101-80-4	5.07	5.09	5.06	5.03	5.06	5.08
18	92-87-5	5.02	5.03	5.03	5.01	5.02	5.01
19	101-77-9	5.08	5.09	5.10	4.99	5.02	4.98
20	97-56-3	5.04	5.08	5.09	5.07	5.09	5.05
21	838-88-0	5.10	5.04	5.07	5.07	5.08	5.04
22	119-93-7	5.03	4.99	5.01	5.04	5.06	5.07
23	139-65-1	5.05	5.02	5.03	5.07	5.08	5.06
24	101-14-4	4.96	5.02	5.05	4.98	4.99	5.03
25	119-90-4	4.98	5.05	5.03	4.99	4.98	4.96
26	91-94-1	5.06	5.08	5.09	5.08	5.09	5.11

Table 4.6.2 Results of robustness study of aromatic amines for GCMS instrument.

Sr. No.	Target Amine	Column oven temperature 100-280 °C	Column oven temperature 95-275 °C	Column oven temperature 105-285 °C	Flow rate, 3.0 mL/min	Flow rate, 3.05 mL/min	Flow rate, 2.95 mL/min
1	62-53-3	4.97	4.94	4.95	5.04	5.06	5.07
2	95-53-4	5.04	5.05	5.03	4.97	4.96	5.06
3	95-68-1	4.97	4.96	4.97	5.05	5.04	5.06
4	87-62-7	4.98	4.97	4.96	5.07	5.08	5.09
5	90-04-0	5.04	5.06	5.03	5.07	5.04	4.94
6	106-47-8	4.97	4.98	4.96	4.95	4.93	4.94
7	106-50-3	5.02	5.08	5.06	5.03	5.02	4.97
8	120-71-8	4.99	4.95	5.06	5.05	5.07	5.06
9	137-17-7	5.04	5.03	5.04	5.04	5.08	5.07
10	95-69-2	5.05	5.06	5.07	5.06	5.09	5.07
11	95-80-7	4.98	5.05	4.99	5.06	5.08	5.05
12	615-05-4	5.06	5.03	5.04	4.97	4.96	5.03
13	91-59-8	5.10	5.04	5.07	5.07	5.08	5.04
14	99-55-8	5.03	4.99	5.01	5.04	5.06	5.07
15	92-67-1	5.05	5.02	5.03	5.07	5.08	5.06
16	60-09-3	5.03	5.05	5.02	4.96	4.95	4.95
17	101-80-4	4.96	5.02	5.05	4.98	4.99	5.03
18	92-87-5	4.98	5.05	5.03	4.99	4.98	4.96
19	101-77-9	5.06	5.08	5.09	5.08	5.09	5.11
20	97-56-3	5.07	5.09	5.06	5.03	5.06	5.08
21	838-88-0	5.02	5.03	5.03	5.01	5.02	5.01
22	119-93-7	5.08	5.09	5.10	4.99	5.02	4.98
23	139-65-1	5.04	5.08	5.09	5.07	5.09	5.05
24	101-14-4	5.08	5.09	5.10	4.99	5.02	4.98
25	119-90-4	5.04	5.08	5.09	5.07	5.09	5.05
26	91-94-1	5.10	5.04	5.07	5.07	5.08	5.04

4.6.1.9 Stability examination: Analyte's stability was examined in the presence of other interference substances in the solution. Analyte's solution was prepared and analysis was performed in the subsequent three days period. Results were interpreted and % age of deviation was calculated for analytes. It was seen that deviation was < 5 %.

Names of 26 aromatic amines, CAS numbers, 2 internal standards and m/z of one target (main) ion and two qualifier (daughter) ions for each amine were summarized in table 4.6.3 & 4.6.4.

Table 4.6.3 Aromatic amines, internal standards (d-8 & d-10) and their CAS numbers.

Sr. No.	Aromatic amine	CAS-No.	Sr. No.	Aromatic amine	CAS-No.
1	Aniline	62-53-3	15	p-cresidine	99-55-8
2	o-toluidine	95-53-4	16	4-aminobiphenyl	92-67-1
3	2,4-xylidine	95-68-1	17	3, 3′-dichlorobenzidine	91-94-1
4	2,6-xylidine	87-62-7	18	1,4-phenylenediamine	60-09-3
5	o-anisidine	90-04-0	19	4-aminophenylether	101-80-4
6	3, 3′-dimethylbenzidine	119-90-4	20	Benzidine	92-87-5
7	4-chloroaniline	106-47-8	21	4, 4′-methylene-bis-benzeneamine	101-77-9
8	4-aminoazobenzene	106-50-3	22	o-aminoazotoluene	97-56-3
9	2, 4-diaminotoluene	120-71-8	23	4, 4′-methylene-bis-(o-toluidine)	838-88-0
10	2, 4, 5-trimethylaniline	137-17-7	24	3, 3′-dimethoxybenzidine	119-93-7
11	4-chloro-2-methylaniline	95-69-2	25	4, 4′-thio-bis-benzenamine	139-65-1
12	2-amino-4-nitrotoluene	95-80-7	26	4, 4′-methylene-bis-(2-chloroaniline)	101-14-4
13	2, 4-diaminoanisole	615-05-4	27	Naphthalene-d8	1146-65-2
14	2-naphthylamine	91-59-8	28	Anthracene-d10	1719-06-8

Table 4.6.4 Aromatic amines selection on GCMS and confirmation of amines from their UV spectrum on UFLC.

Sr. No.	CAS. No.	Type	m/z Target ion on GC	Qualifier one m/z on GC	Qualifier two m/z on GC	UV visible Maxima for identification (nm) on UFLC	Wavelength for quantification (nm) on UFLC
1	62-53-3	Target	93	66	65	232.6, 283.4	240
2	95-53-4	Target	106	107	77	232.6, 283.4	240
3	95-68-1	Target	121	120	106	232.6, 281.0	240
4	87-62-7	Target	121	106	120	233.8, 288.1	240
5	90-04-0	Target	108	80	123	233.8, 284.5	240
6	1146-65-2	Internal Standard	136	-	-	-	-
7	106-47-8	Target	127	129	92	239.7, 294.0	240
8	106-50-3	Target	108	80	107	240.8, 305.9	240
9	120-71-8	Target	122	94	137	236.1, 289.3	240
10	137-17-7	Target	120	135	134	233.8, 289.3	240
11	95-69-2	Target	141	106	140	240.8, 292.8	240
12	95-80-7	Target	122	121	94	238.5, 295.2	240
13	615-05-4	Target	123	138	95	242, 302.3	240
14	91-59-8	Target	143	115	116	236.1, 281.0	240
15	99-55-8	Target	152	77	106	250.3, 296.4	240
16	92-67-1	Target	169	168	141	275.1	280
17	1719-06-8	Internal Standard	188	-	-	-	-
18	60-09-3	Target	92	197	120	247.9, 382.0	240
19	101-80-4	Target	200	108	71	244.4, 297.6	240
20	92-87-5	Target	184	185	156	282.2	280
21	101-77-9	Target	198	197	106	243.2, 288.1	240
22	97-56-3	Target	106	225	91	255.0, 383.2	240
23	838-88-0	Target	226	211	120	242.0, 289.3	240
24	119-93-7	Target	212	180	106	282.2	280
25	139-65-1	Target	216	184	80	263.2	240
26	101-14-4	Target	231	266	268	245.6, 295.2	240
27	119-90-4	Target	244	201	229	282.2, 305.9	305
28	91-94-1	Target	252	254	253	285.7	280

Each amine was registered against the m/z of one target ion and two qualifiers' ions in the GCMS instrument library. When analyzing the processed test specimens on GCMS instrument, it detects the amines by matching their m/z spectrum to the saved library (as shown in Fig. 4.6.1 and 4.6.2). We draw a calibration curve and save in the instrument. Processed test specimens were programed against this calibration curve. Instrument run the programmed test specimens and quantify the results by matching the signals to the curve points (as shown in Fig. 4.6.0).

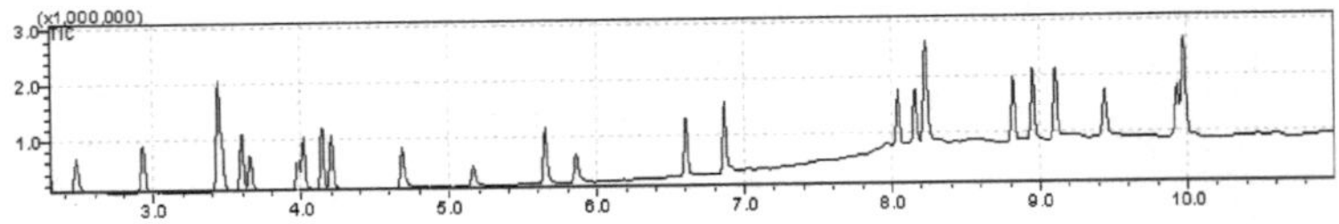

Fig. 4.6.1 Amines peak profile from GC-MS Instrument

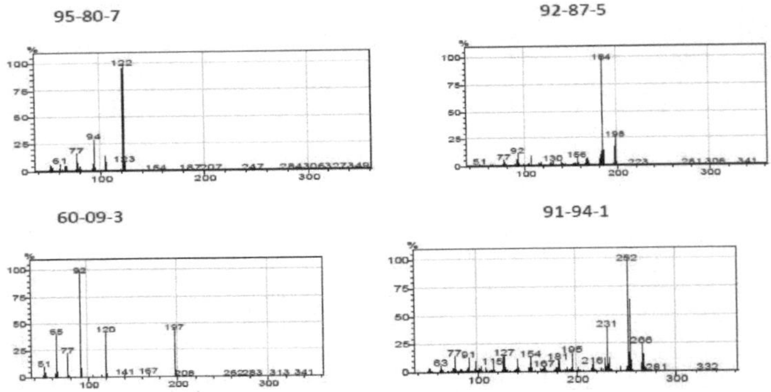

Fig. 4.6.2 GC-MS Mass spectrums of some mostly detected amines (Data from instrument)

In the UFLC instrument each amine was conformed from its UV spectrum. Spectrum of each amine has its own specific shape, retention time, maxima and minima (that were given in table 4.6.4). Standard amines were analyzed on UFLC instrument and spectrum of each amine was saved in the instrument library. Each detected amine from GCMS instrument was conformed in the UFLC by matching its spectrum to the standard amine spectrum save in the UFLC instrument library (as shown in Fig. 4.6.3 and 4.6.4). Test specimens were performed using ISO methods and newly established method and results were compared. Results obtained against both methods were comparable and there was not much difference between the results (as shown in table 4.6.5). Hence it was concluded from the results that the newly established method can be used for further analysis of test specimens.

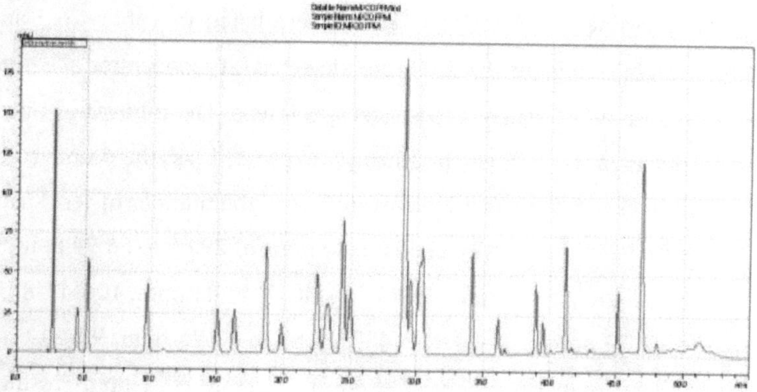

Fig. 4.6.3 UFLC Instrument peak profile data for aromatic amines

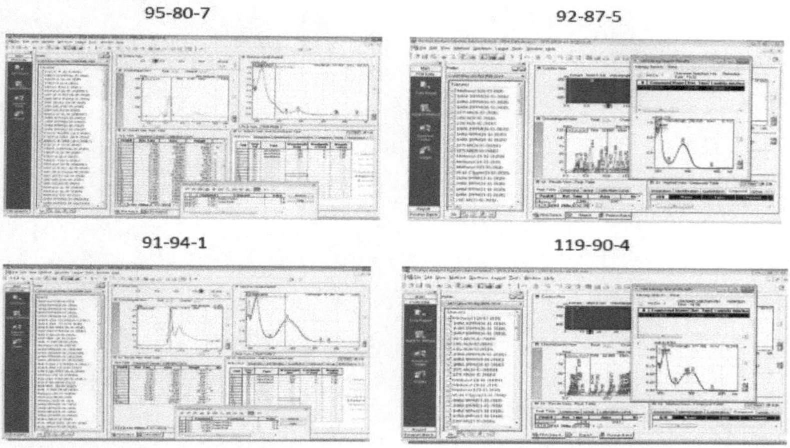

Fig. 4.6.4 UFLC Instrument UV spectrums of some aromatic amines, that were compared with standard spectrums

4.6.2 Concentrations of aromatic amines in the natural and printed fibers

Four hundred eighty-seven textile test specimens made of natural fibers as cotton, wool and silk fibers were analyzed without xylene extraction. The frequency of detection and the concentration range of each detected aromatic amine in the test specimens were listed in Table 4.6.5 and Fig. 4.6.5. Eight kinds of PAAs were detected at concentrations over the reporting limit (≥ 5 ppm) in these test specimens. The following compounds were detected at a high frequency (expressed as the number of test specimens detected over the reporting limit / the number of test specimens analyzed): 92-87-5 (4/487), at concentrations of 29.66 to 56.94 ppm, 92-67-1 (2/487), 5.18 to 8.54 ppm, 91-94-1 (1/487), 13.41 ppm, 106-47-8 (3/487), 9.25 to 21.29 ppm, 119-90-4 (3/487), 5.58 to 13.85 ppm, 90-04-0 (2/487) 9.28 to 19.6 ppm, and 95-80-7 (2/487), 37.19 to 52.2 ppm (results were summarized in table 4.6.5). 5/487 test specimens contain amines above EU regulations (≥ 30 ppm). It is thought that these textile products should be used more carefully.

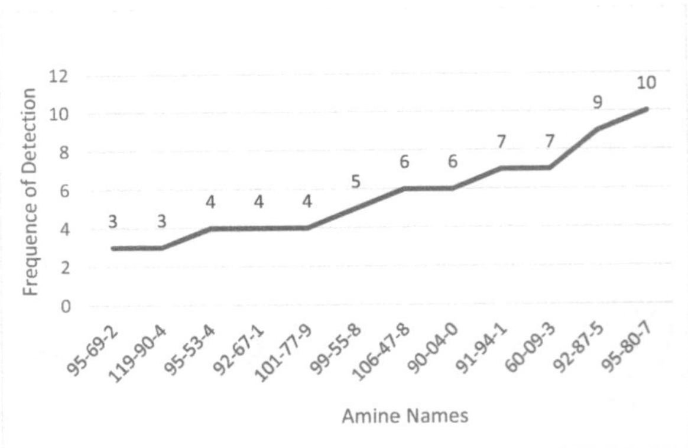

Fig. 4.6.5 Frequency of detection of amines in the three-month period

4.6.3 Concentrations of aromatic amines in the synthetic fibers

Three thirty-two test specimens which contain synthetic fibers such as polyester and nylon were analyzed by means of solvent extraction using xylene. Five kinds of aromatic amines were detected at concentrations that exceeded the reporting limit in these test specimens. The following amines were detected at a high frequency: 60-09-3 (3/332), in the range 62.07 to 188.55 ppm, 92-87-5 (1/332), up to 38.02 ppm, 95-80-7 (3/332), 5.38 to 37.19 ppm, 90-04-0 (2/332), 19.6 to 75.38 ppm and 106-47-8 (2/332), 15.57 to 21.29 ppm. 8/332 test specimens contain amines above EU regulations ≥ 30 ppm. Concentrations of other amines were also present below the reporting limit (results were summarized in table 4.6.5).

4.6.4 Concentrations of aromatic amines in the liquid and powder dyes

Thirty-four test specimens of liquid and powder dyes were analyzed. Five kinds of aromatic amines were detected at concentrations that exceeded the reporting limit in these test specimens. 92-67-1 (3/34), in the range 53.78 to 376.46 ppm, 120-71-8 (3/34), 34.83 to 415.43 ppm, 92-87-5 (2/34), 66.11 to 74.87 ppm, 95-53-4 (1/34), 39.42 ppm and 90-04-0 (3/34), 241.58 to 1798.14 ppm. 8/34 test specimens contain amines above reporting limit ≥ 25 ppm and 6/34 amines were above EU regulations limit for liquid and powder dyes ≥ 200 ppm (results were summarized in table 4.6.5).

4.6.5 Concentrations of aromatic amines in the Leather test specimens

Fourteen test specimens of leather were analyzed. Two kinds of aromatic amines were detected at concentrations that exceeded the reporting limit in these test specimens. 92-67-1 (1/14), up to 8.65 ppm, and 92-87-5 (2/14), 41.2 to 632.94 ppm. 3/14 test specimens contain amines above reporting limit ≥ 5 ppm and 2/14 test specimens contain amines above EU regulations ≥ 30 ppm (results were summarized in table 4.6.5).

Table 4.6.5 Comparison of results of detectable test specimens analyzed on GCMS and UFLC by ISO methods and newly established method

Sr. No.	Test specimen description	Color	Composition and Type applied	CAS No. for Amine Detected	GCMS Results (ISO Methods) ppm[a]	GCMS Results (New method) ppm[b]	UFLC Results	Mean [a+b÷2]	Standard Deviation (SD)	%Relative Standard
1	Loose Fabric	Yellow	Cotton (A)	92-87-5	53.68	60.20	52.02	56.94	4.61	8.10
2	Loose Fabric	Pink	Cotton (A)	92-67-1	8.52	8.55	11.11	8.54	0.02	0.25
3	Printed Garment (Hood)	Brown Print	Cotton Printed (N/A)	91-94-1	13.65	13.16	10.74	13.41	0.35	2.58
4	Finished Garment (printed)	Black and blue color print	Cotton Printed (N/A)	92-87-5 106-47-8	29.66 9.51	27.76 8.34	36.96 7.89	28.71 8.93	1.34 0.83	4.68 9.27
5	Finished Garment (Shirt)	Violet	Cotton (A)	92-87-5 119-90-4	40.66 5.66	39.89 5.50	37.66 5.56	40.28 5.58	0.54 0.11	1.35 2.03
6	Loose Fabric	Yellow	Cotton (A)	92-87-5 92-67-1	36.63 5.10	34.00 5.25	34.63 5.10	35.32 5.18	1.86 0.11	5.27 2.05
7	Gloves	Blue	Cotton (A)	119-90-4	6.68	8.20	9.50	7.44	1.07	14.45
8	Loose Fabric	Pink	Cotton (A)	119-90-4	13.13	14.57	16.46	13.85	1.02	7.35
9	Finished Garment (Printed)	Purple, pink and black print	Cotton Printed with Natural dye (N/A)	90-04-0 106-47-8 95-80-7	9.26 9.94 54.06	9.30 8.56 50.34	14.26 10.77 64.01	9.28 9.25 52.20	0.03 0.98 2.63	0.30 10.55 5.04
10	Finished Garment (Printed)	Black and blue print	Cotton (N/A)	90-04-0 106-47-8 95-80-7	18.09 19.93 36.58	21.10 22.64 37.79	22.80 24.67 35.71	19.60 21.29 37.19	2.13 1.92 0.86	10.86 9.00 2.30
11	Gloves	Black and blue	Synthetic fiber (C)	95-80-7	5.64	5.11	6.78	5.38	0.37	6.97
12	Finished Garment (Shirt)	Brown	Cotton polyester (C)	60-09-3	89.76	87.56	92.45	88.66	1.56	1.75
13	Finished Garment (Female Shirt)	Purple	Cotton polyester (C)	92-87-5	37.03	39.02	37.03	38.02	1.41	3.70
14	Finished Garment (Hood)	Composite of three colors (black, blue and pink)	Cotton polyester (C)	60-09-3	60.64	63.50	79.12	62.07	2.02	3.26
15	Printed Garment (Pant)	Red and black print	Cotton polyester (C)	90-04-0 106-47-8 95-80-7	72.20 16.37 45.56	78.56 14.76 43.37	71.33 18.23 47.55	75.38 15.57 44.47	4.50 1.14 1.55	5.97 7.31 3.48
16	(Gloves)	Brown	Synthetic fiber (C)	90-04-0 106-47-8 95-80-7	14.20 23.60 47.70	11.80 26.60 45.4	13.00 24.40 49.40	19.60 21.29 37.19	2.13 1.92 0.86	10.86 9.00 2.30

17	Loose Fabric	Red	Polyester (B)	60-09-3	191.78	185.32	188.23	188.55	4.57	2.42
18	Loose Fabric	Brown	Polyester (B)	60-09-3	238.53	227.74	255.46	233.14	7.63	3.27
19	Leather Jacket	Brown	Leather	92-67-1 92-87-5	8.90 632.82	8.40 633.05	10.21 616.34	8.65 632.94	0.35 0.16	4.09 0.03
20	Leather Shoes	Black	Leather	92-87-5	39.34	43.05	39.34	41.20	2.62	6.37
21	Liquid Dye	Black	Liquid dye (N/A)	90-04-0	1796.32	1799.96	1800.85	1798.14	2.57	0.14
22	Powder Dye	Brown	Powder dye (N/A)	92-67-1	382.90	370.02	342.11	376.46	9.18	2.42
23	Powder Dye	Black	Powder dye (N/A)	120-71-8	416.10	414.75	422.21	415.43	0.95	1.95
24	Powder Dye	Yellow	Powder dye (N/A)	92-67-1 92-87-5 120-71-8	132.36 175.70 36.62	136.06 174.03 33.04	129.62 168.98 40.97	134.21 174.87 34.83	2.62 1.18 2.53	1.95 0.68 7.27
25	Powder Dye	Brown	Powder dye (N/A)	92-67-1 92-87-5 120-71-8	52.46 64.88 284.45	55.10 67.34 289.66	48.11 60.98 302.42	53.78 66.11 287.06	1.87 1.74 3.68	3.47 2.63 1.28
26	Powder Dye	Pink	Powder dye (N/A)	95-53-4	40.07	38.76	43.35	39.42	0.93	2.35
27	Powder Dye	Red	Powder dye (N/A)	90-04-0	244.73	238.42	238.42	241.58	4.46	1.85
28	Powder Dye	Yellow	Powder dye (N/A)	90-04-0	378.56	385.82	282.22	382.19	5.13	1.34

Type A= Non-extraction (Natural method),

Type B= Extraction method (color extraction with xylene), Type C= Combine Method,

Type N/A = Type not applicable

THESIS

CONCLUSION

Conclusion:

For quantitative determination of chemical toxicants from the matrix as leather, foam, textile and plastic samples new, rapid and sensitive methods were developed. Adequate solvents were used for the extraction of target analytes from respective matrix. Multiple types of toxicants as regulated flame retardants, phthalates, biocide contents (MIT, CIT, OIT and Triclosan), Ortho-phenylphenols, DMFu and azo dyes (aromatic amines) were determined quantitatively. These toxicants were extracted and examined from respective matrix using newly established analytical methods. Suitability of the solvent extraction process was examined before examining the real test specimens against any established analytical method. Sonication method and suitable organic solvents were utilized for the extraction of target analytes from the matrix. Suitable columns were selected for analyzing the analytes. Under the optimized conditions the target analytes present linearity in the range ≥ 0.994 to 0.999. Subjected target analytes relative standard deviation was < 10 %. Recovery was in the range 90% to 110 %. These subjected analytical methods were very specific, sensitive and useful tools for examining the safety of leather and textile test specimens. These analytical methods enable the successful examination of target analytes from multiple types of matrices, so the human exposure of these toxicants can be tracked successfully.

REFRENCES

References

[1] Y. Zhang, H. Su, M. Ya, J. Li, S.-H. Ho, L. Zhao, K. Jian, R. J. Letcher and G. Su, Science of The Total Environment **693**, 133654 (2019).

[2] T. T. Yang, J.-P. Guan, R.-C. Tang and G. Chen, Industrial Crops and Products **115**, 16 (2018).

[3] A. K. Salmeia, S. Gaan and G. Malucelli, Polymers **8** (9) 319 (2016).

[4] C. Liu, T. Xing, B. Wei and G. Chen, Materials **11** (10) 1842 (2018).

[5] S. C. Hammel, K. Hoffman, A. M. Lorenzo, A. Chen, A. L. Phillips, C. M. Butt, J. A. Sosa, T. F. Webster and H. M. Stapleton, Environment International **107**, 181 (2017).

[6] M. Brits, S. H. Brandsma, E. R. Rohwer, J. De Vos, J. M. Weiss and J. de Boer, Environmental Pollution **253**, 120 (2019).

[7] A. de la Torre, I. Navarro, P. Sanz and M. de los Ángeles Martínez, Journal of Hazardous Materials **382**, 121009 (2020).

[8] C. F. Estill, J. Slone, A. C. Mayer, K. Phillips, J. Lu, I. C. Chen, A. Christianson, R. Streicher, M. J. L. Guardia, N. Jayatilaka, M. Ospina and A. M. Calafat, Journal of Occupational and Environmental Hygiene **16** (7), 477 (2019).

[9] Y. J. Hu, L. J. Bao, C. L. Huang, S. M. Li and E. Y. Zeng, Environmental Pollution **252**, 1902 (2019).

[10] K. Jarova, A. Koleckarova, Z. Kralova and M. Vavrova, Neuro Endocrinol Lett **37** (Suppl1), 45 (2016).

[11] G. Rivière, J. Jean, S. Gorecki, M. Hulin, M. Kolf-Clauw, C. Feidt, N. Picard-Hagen, P. Vasseur, B. Le Bizec and V. Sirot, Food and Chemical Toxicology **131**, 110561 (2019).

[12] M. J. Piazza and A. A. Urbanetz, in *JBRA Assist Reprod* **23**, 154 (2019)

[13] Y. Liang, X. Liu and M. R. Allen, Environmental Science & Technology **52** (10), 5821 (2018).

[14] A. M. Al-Salem, Q. Saquib, M. A. Siddiqui, J. Ahmad, R. Wahab and A. A. Al-Khedhairy, Chemosphere **237**, 124519 (2019).

[15] H. Kang, J. Lee, J. P. Lee and K. Choi, Environment International **131**, 105034 (2019).

[16] D. Nacci, B. Clark, M. J. La Guardia, K. Miller, D. Champlin, I. Kirby, A. Bertrand and S. Jayaraman, Environmental Toxicology and Chemistry **37** (9), 2350 (2018).

[17] S. Yuan, Y. Han, M. Ma, K. Rao, Z. Wang, R. Yang, Y. Liu and X. Zhou, Environmental Pollution **250**, 58 (2019).

[18] N. Wemken, D. S. Drage, C. Cellarius, K. Cleere, J. J. Morrison, S. Daly, M. A.-E. Abdallah, C. Tlustos, S. Harrad and M. A. Coggins, Environmental Research **180**, 108805 (2020).

[19] M. E. Street, S. Angelini, S. Bernasconi, E. Burgio, A. Cassio, C. Catellani, F. Cirillo, A. Deodati, E. Fabbrizi, V. Fanos, G. Gargano, E. Grossi, L. Iughetti, P. Lazzeroni, A. Mantovani, L. Migliore, P. Palanza, G. Panzica, A. M. Papini,

S. Parmigiani, B. Predieri, C. Sartori, G. Tridenti and S. Amarri, International Journal of Molecular Sciences **19** (6), 1647 (2018).

[20] F. Iadaresta, M. D. Manniello, C. Östman, C. Crescenzi, J. Holmbäck and P. Russo, Environmental Science and Pollution Research **25** (25), 24629 (2018).

[21] N. K. Jayatilaka, P. Restrepo, Z. Davis, M. Vidal, A. M. Calafat and M. Ospina, Chemosphere **235**, 481 (2019).

[22] D. W. Killilea, D. Chow, S. Q. Xiao, C. Li and M. L. Stoller, Toxicology Reports **4**, 260 (2017).

[23] N. Li, W. Ho, R. S. Sun Wu, G.-G. Ying, Z. Wang, K. Jones and W.-J. Deng, Ecotoxicology and Environmental Safety **183**, 109502 (2019).

[24] L. Hu, Y. Tao, D. Luo, J. Feng, L. Wang, M. Yu, Y. Li, A. Covaci and S. Mei, Chemosphere **233**, 724 (2019).

[25] P. Krystek, H. Beeltje, M. Noteboom, E. M. van den Hoeven and M. M. G. Houtzager, Journal of Pharmaceutical and Biomedical Analysis **170**, 169 (2019).

[26] Q. Luo, S. Wang, M. Adeel, Y. Shan, H. Wang and L.-n. Sun, Scientific Reports **9** (1), 11292 (2019).

[27] Y. Miyake, M. Tokumura, H. Nakayama, Q. Wang, T. Amagai, S. Ogo, K. Kume, T. Kobayashi, S. Takasu, K. Ogawa and K. Kannan, Science of The Total Environment **601-602**, 1333 (2017).

[28] D. Zacs, I. Perkons, V. Volkovs and V. Bartkevics, Chemosphere **230**, 396 (2019).

[29] C. Charron, S. De Vaugelade, F. Richard, A. Largitte and S. Pirnay, International Journal of Cosmetic Science **40** (3), 269 (2018).

[30] F. Buiarelli, P. D. Filippo, D. Pomata, C. Riccardi and M. Bartocci, Int. J. Of Environ. Anal. Chem., **97**(9), 797 (2017).

[31] D. Pomata, P. D. Filippo, C. Riccardi, V. Rossi, G. Simonetti, E. Sonego and F. Buiarelli, Int. J. Of Environ. Anal. Chem., (2019). https://doi.org/10.1080/03067319.2019.1655007;

[32] W. A. Stubbings, L.V. Nguyen, K. Romanak, L. Jantunen, L. Melymuk, V. Arrandale, M.L. Diamond and M. Venier, Sci Total Environ., **675**, 594 (2019).

[33] H. Yanagisawa, Y. Kudo, K. Nakagawa, H. Miyagawa, F. Maruyama and S. Fujimaki, Molecules **23** (4), 728 (2018).

[34] C. Warembourg, X. Basagaña, C. Seminati, J. de Bont, B. Granum, S. Lyon-Caen, C.B. Manzano-Salgado, I. Pin, A.K. Sakhi, V. Siroux, R. Slama, J. Urquiza, M. Vrijheid, C. Thomsen and M. Casas, Int. J. Hyg. Environ. Health, **222,** 446 (2018).

[35] M.E. Street, S. Angelini, S. Bernasconi, E. Burgio, A. Cassio, C. Catellani, F. Cirillo, A. Deodati, E. Fabbrizi, V. Fanos, G. Gargano, E. Grossi, L. Iughetti, P. Lazzeroni, A. Mantovani, L. Migliore, P. Palanza, G. Panzica, A.M. Papini, S. Parmigiani, B. Predieri, C. Sartori, G. Tridenti and S. Amarri, Int. J. Mol. Sci. **19** (2018). doi:10.3390/ijms19061647

[36] R. S. Strakovsky and S.L. Schantz, Environ. Epigenet. **4** (3) (2018). doi:10.1093/eep/dvy022.

[37] Y.Q. Ren, G.H. Wang, J.J. Li, C. Wu, C. Cao, J. Li, J.Y. Wang, S.S. Ge, Y.N. Xie, X.R. Li, F. Meng and H. Li, Chemosphere **215**, 413 (2019). doi:10.1016/j.chemosphere.2018.10.064.

[38] D.L. McBride, J. Pediatri. Nurs. **45,** 76 (2018).

[39] N. Bandow, S. Gartiser, O. Ilvonen and U.J.E.S.E. Schoknecht, Environ. Sci. Eur. **30,** 14 (2018).

[40] R.M. Martinez, R. Hauser, L. Liang, A. Mansur, M. Adir, L. Dioni, C. Racowsky, V. Bollati, A.A. Baccarelli and R. Machtinger, Environ. Int. **123,** 20 (2019). doi:10.1016/j.envint.2018.11.043.

[41] E. Cullen, D. Evans, C. Griffin, P. Burke, R. Mannion, D. Burns, A. Flanagan, A. Kellegher, G. Schoeters, E. Govarts, P. Biot, L. Casteleyn, A. Castaño, M. Kolossa-Gehring, M. Esteban, G. Schwedler, H.M. Koch, J. Angerer, L.E. Knudsen, R. Joas, A. Joas, B. Dumez, O. Sepai, K. Exley and D. Aerts, Int. J. Environ. Res. Public Health **14** (12), 1456 (2017).

[42] R.M. Shaffer, K.K. Ferguson, L. Sheppard, T. James-Todd, S. Butts, S. Chandrasekaran, S.H. Swan, E.S. Barrett, R. Nguyen, N. Bush, T.F. McElrath and S. Sathyanarayana, Environ. Int. **123,** 588 (2019). doi:10.1016/j.envint.2018.12.021.

[43] C. Ursino, R. Castro-Muñoz, E. Drioli, L. Gzara, M.H. Albeirutty and A. Figoli, Membranes **8** (2), 18 (2018).

[44] L. Cheng, S. Pan, C. Ding, J. He and C. Wang, J. Chromatogr. A **1511**, 85 (2017). doi:10.1016/j. chroma.2017.07.012.

[45] C. Wang, J. Sep. Sci. **42**, 1248–1256-2019 v.1242 no.1246 (2019). doi:10.1002/jssc.201801171.

[46] M. Long, M. Ghisari, L. Kjeldsen, M. Wielsøe, B. Nørgaard-Pedersen, E.L. Mortensen, M.W. Abdallah and E.C.J.M.A. Bonefeld-Jørgensen, Mol. Autism **10** (1), 1 (2019).

[47] K. Minjeong, J. Ji Seong, K. Hyunji, H. Seungwoo, P. Il-Hyun, L. Byung-Chul, Y. Sung Il, J.S. Ha, N.K. Taek and L. Kyung-Min, Biomol. Ther. (Seoul) **26,** 512 (2018). doi:10.4062/ biomolther.2018.076.

[48] REACH. Website (new Banned Phthalates Included in RSL List 2018). (2018). https://eur-lex. europa.eu/eli/reg/2018/2005/oj

[49] S. Hartmann and U.J.E.S.E. Klaschka, PLoS One **30**, 29 (2018).

[50] S. Qian, H. Ji, X. Wu, N. Li, Y. Yang, J. Bu, X. Zhang, L. Qiao, H. Yu, N. Xu and C. Zhang, PLoS ONE **13** (12), e0208467 (2018).

[51] Y.-L. Fan, M. Ma, H.-Z. Cui, S. Liu, M.-L. Yang and X.-F. Hou, J. Sep. Sci. **41,** 2993 (2018).

[52] C. Charron, S. De Vaugelade, F. Richard, A. Largitte and S. Pirnay, J. Cosmet. Sci. **40,** 269 (2018).

[53] J. Lin, W. Chen, H. Zhu and C. Wang, J. Dairy Sci. **98,** 8278 (2015). doi:10.3168/jds.2015- 10066.

[54] Council Regulation (EU) (2017) 2017/997 amending Annex III to Directive 2008/98/EC as regards the hazardous property HP 14 'Ecotoxic' https://eur-lex.europa.eu/legal-content/EN/TXT/?uri=CELEX%3A32017R0997)

[55] Regulation (EU) No 528/2012 of the European Parliament and of the Council of 22 May 2012 concerning the making available on the market and

use of biocidal products. https://eur-lex.europa.eu/LexUriServ/LexUriServ.do?uri=OJ:L:2012:167:0001:0123:EN. PDF

[56] Emission scenario documents for product type 9, https://echa.europa.eu/de/guidance-documents/guidance-on-biocides-legislation/emission-scenario-documents

[57] D. P. Bockmühl, J. Schages and L. Rehberg, Microbial cell (Graz, Austria), **6** (7), 299 (2019). doi:10.15698/mic2019.07.682. (Accession No. 31294042)

[58] N. Hamnerius, A. Pontén and M. Mowitz, Contact dermatitis, **78** (6), 419 (2018). doi:10.1111/cod.12959

[59] D. Höfer, Dermatology research and practice, 2018, 7043438 (2018). doi:10.1155/2018/7043438. (Accession No. 30581460)

[60] K. S. Huang, C. H. Yang, S. L. Huang, C. Y. Chen, Y.Y. Lu and Y. S. Lin, International Journal of Molecular Sciences, **17** (9) (2016). doi:10.3390/ijms17091578. (Accession No. 27657043)

[61] F. Iadaresta, M. D. Manniello, C. Östman, C. Crescenzi, J. Holmbäck and P. Russo, Environmental science and pollution research international, **25** (25), 24629 (2018). doi:10.1007/s11356-018-2448-6

[62] P. J. Rivero, A. Urrutia, J. Goicoechea and F. J. Arregui, Nanoscale research letters, **10** (1), 501 (2015). doi:10.1186/s11671-015-1195-6

[63] U. Schoknecht, H. Mathies and R. Wegner, Environmental sciences Europe, **28** (1), 6 (2016). doi:10.1186/s12302-016-0074-9

[64] T. Walser, E. Demou, D. J. Lang and S. Hellweg, Environmental Science & Technology, **45** (10), 4570 (2011). doi:10.1021/es2001248

[65] D. S. Morais, R. M. Guedes and M. A. Lopes, Materials (Basel, Switzerland), **9** (6) (2016). doi:10.3390/ma9060498. (Accession No. 28773619)

[66] Z. F. Chen, G. G. Ying, H. J. Lai, F. Chen, H. C. Su, Y. S. Liu and J. L. Zhao, Analytical and bioanalytical chemistry, 404 (2012). doi:10.1007/s00216-012-6444-2

[67] N. Bandow, S. Gartiser, O. Ilvonen and U. Schoknecht, Environmental sciences Europe, **30** (1), 14 (2018). doi:10.1186/s12302-018-0144-2

[68] J. J. Heo, U. J. Kim and J. E. Oh, Environmental Engineering Research, **24** (1), 137 (2019). doi:10.4491/eer.2018.144

[69] B. T. Li, Z. Chen, W. L. Wang, Y. X. Sun, T. H. Zhou, A. Li and H. Y. Hu, Water, **10** (4), 532 (2018). https://www.mdpi.com/2073-4441/10/4/532

[70] H. Singer, S. Jaus, I. Hanke, A. Lück, J. Hollender and A. C. Alder, (2010). Environmental Pollution, **158** (10), 3054 (2010). doi:https://doi.org/10.1016/j.envpol.2010.06.013

[71] N. Voulvoulis, M. D. Scrimshaw and J. N. Lester, Chromatographia, 50(5), 353 (1999). doi:10.1007/BF02490841

[72] C. Wang, T. Xie, R. Xu, J. Lin, L. Li, C. W. S. E. E. Inspection and Q. Bureau, World Journal of Engineering and Technology, **5** (1) (2017). Retrieved from https://www.scilit.net/articlebfa2b4a0ba24ed8b7a5215236 e1565cb

[73] V. Giannone, I. Pitino, B. Pecorino, A. Todaro, A. Spina, M. R. Lauro, F. Tomaselli and C. Restuccia, Int J Food Microbiol **235**, 71-76 (2016).

[74] H. Kuwahara, J. Ninomiya and H. Morita, Biocontrol Science **23** (3), 85-96 (2018).

[75] K. S. Jewell, A. Wick and T. A. Ternes, Water Research **48**, 478-489 (2014).

[76] P. Karas, A. Metsoviti, V. Zisis, C. Ehaliotis, M. Omirou, E. S. Papadopoulou, U. Menkissoglou-Spiroudi, S. Manta, D. Komiotis and D. G. Karpouzas, Sci Total Environ **530-531**, 129-139 (2015).

[77] P. A. Karas, S. Makri, E. S. Papadopoulou, C. Ehaliotis, U. Menkissoglu-Spiroudi and D. G. Karpouzas, Ecotoxicol Environ Saf **124**, 447-454 (2016).

[78] C. Perruchon, V. Patsioura, S. Vasileiadis and D. G. Karpouzas, Pest management science **72** (1), 113-124 (2016).

[79] N. Hosoya, K. Motomura, E. Tagawa, C. Ogihara, M. Nagano and H. Hosoya, *Effects of the fungicide ortho-phenylphenol (OPP) on the early development of sea urchin eggs*. Mar Environ Res **143**, 24-29 (2018).

[80] E. Nixon, J. D. Brooks, P. A. Routh, J. T. Chittenden and R. E. Baynes, Journal of applied toxicology: JAT **37** (4), 508-512 (2017).

[81] M. Olak-Kucharczyk and S. Ledakowicz, Journal of Hazardous Materials **333**, 348-357 (2017).

[82] C. Perruchon, S. Vasileiadis, C. Rousidou, P. Evangelia, G. Tanou, M. Samiotaki, C. Garagounis, A. Molassiotyis, K. Papadopoulou and D. Karpouzas, Author Correction: Metabolic pathway and cell adaptation mechanisms revealed through genomic, proteomic and transcription analysis of a Sphingomonas haloaromaticamans strain degrading ortho-phenylphenol. Scientific reports **8**, 4599 (2018).

[83] C. Perruchon, S. Vasileiadis, C. Rousidou, E. S. Papadopoulou, G. Tanou, M. Samiotaki, C. Garagounis, A. Molassiotis, K. K. Papadopoulou and D. G. Karpouzas, Scientific reports **7** (1), 6449-6449 (2017).

[84] S. Balakrishnan and D. A. Eastmond, Food and Chemical Toxicology **44** (8), 1340-1347 (2006).

[85] D. Brusick, Environmental and molecular mutagenesis **45** (5), 460-481 (2005).

[86] A. Hartwig, *ortho-Phenylphenol (OPP) und ortho-Phenylphenol-Natrium (OPP-Na) [MAK Value Documentation in German language, 2016]*. (2016).

[87] H. J. Jang, C. Nde, F. Toghrol and W. E. Bentley, BMC genomics **9**, 411-411 (2008).

[88] C. W. Nde, H.-J. Jang, F. Toghrol and W. E. Bentley, BMC genomics **9**, 473-473 (2008).

[89] M. Coelhan, K.-H. Bromig, K. Glas and A. L. Roberts, Journal of Agricultural and Food Chemistry **54** (16), 5731-5735 (2006).

[90] M. Coelhan, J. T. Yu and A. L. Roberts, Food Chemistry **112** (2), 515-519 (2009).

[91] B. Saad, N. Hana Haniff, M. Saleh, N. Hashim, A. Abu and N. Ali, Determination of ortho-phenylphenol, diphenyl and diphenylamine in apples and oranges using HPLC with fluorescence detection. Food Chemistry **84** (2), 313-317 (2004).

[92] L. Yang, A. Kotani, H. Hakamata and F. Kusu, Anal Sci **20** (1), 199-203 (2004).

[93] M. Wannick, J. C. Assmann, J. F. Vielhauer, S. Offermanns, D. Zillikens, C. D. Sadik and M. Schwaninger, Front Immunol **9** 1890 (2018).

[94] P. Perini, F. Rinaldi, M. Puthenparampil, M. Marcon, F. Perini and P. Gallo, Mult Scler Relat Disord **26**, 68 (2018).

[95] J. L. Andersen, B. Gesser, E. D. Funder, C. J. F. Nielsen, H. Gotfred-Rasmussen, M. K. Rasmussen, R. Toth, K. V. Gothelf, J. S. C. Arthur, L. Iversen and P. Nissen, Nature Communications **9**, 4344 (2018).

[96] A. A. Ahmed, A. A. E. Ahmed, E. M. El Morsy and S. Nofal, J Pharm Pharmacol **70**, 1521 (2018).

[97] D. Ontaneda, J. Nicholas, M. Carraro, J. Zhou, Q. Hou, J. Babb, K. Riester, J. P. Mendoza, T. Livingston and M. Jhaveri, Mult Scler Relat Disord **27**, 101 (2018).

[98] L. Liu, M. K. Vollmer, A. S. Ahmad, V. M. Fernandez, H. Kim and S. Doré, Free Radical Biology and Medicine **131**, 98 (2019).

[99] P. Stefanelli, S. Girolimetti, A. Santilio, R. Dommarco, Bull Environ Contam Toxicol **86**, 428 (2011).

[100] E. D'Amico, A. Zanghì, G. Callari, G. Borriello, A. Gallo, G. Graziano, P. Valentino, M. Buccafusca, S. Cottone, G. Salemi, P. Ragonese, R. B. Bossio, R. Docimo, L. M. E. Grimaldi, C. Pozzilli, G. Tedeschi, M. Zappia and F. Patti, in *Ther Adv Neurol Disord* (2018), Vol. 11, pp. 1756286418796404.

[101] E. D'Amico, A. Zanghì, M. Sciandra, G. Borriello, G. Callari, A. Gallo, G. Salemi, S. Cottone, M. Buccafusca, P. Valentino, R. B. Bossio, L. M. E. Grimaldi, C. Pozzilli, G. Tedeschi, M. Zappia and F. Patti, J Neurol **266**, 411 (2018).

[102] S. Condé, X. Moisset, B. Pereira, M. Zuel, R. Colamarino, M. Maillet-Vioud, M. Lauxerois, F. Taithe and P. Clavelou, Eur J Neurol **26**, 460 (2018).

[103] S. Braune, S. Grimm, P. van Hövell, U. Freudensprung, F. Pellegrini, R. Hyde, A. Bergmann J. Neurology, 265, 2980 (2018).

[104] Commission regulation No. 412/2012, official journal of the European union 16 may 2012 **128**, 1 (2012).

[105] R. S. Abdelrahman and N. Abdel-Rahman, Life Sciences **217**, 251 (2019).

[106] T. J. Braley, A. K. Huber, B. M. Segal, N. Kaplish, R. Saban, J. M. Washnock-Schmid and R. D. Chervin, Sleep **41**, 109 (2018).

[107] A. A. Abd El-Fattah, A. T. Fahim, N. A. H. Sadik and B. M. Ali, Brain Research **1701**, 227 (2018).

[108] N. Singh, L. Saha, P. Kumari, J. Singh, A. Bhatia, D. Banerjee and A. Chakrabarti, Brain Research Bulletin **144**, 233 (2019).

[109] R. A. Høglund, J. Polak, F. Vartdal, T. Holmøy and A. Lossius, Mult Scler Relat Disord **26**, 90 (2018).

[110] M. Nachliely, A. Trachtenberg, B. Khalfin, K. Nalbandyan, M. Cohen-Lahav, K. Yasuda, T. Sakaki, A. Kutner and M. Danilenko, The Journal of Steroid Biochemistry and Molecular Biology **188**, 8 (2018).

[111] R. Motterlini, A. Nikam, S. Manin, A. Ollivier, J. L. Wilson, S. Djouadi, L. Muchova, T. Martens, M. Rivard and R. Foresti, Redox Biology **20**, 334 (2019).

[112] J. P. Lamas, L. Sanchez-Prado, C. Garcia-Jares and M. Llompart, Journal of Chromatography A **1216**, 5755 (2009).

[113] Y. Lu and Y. Zhu, Talanta **119**, 430 (2014).

[114] R. Narizzano, F. Risso, G. Venturelli, C. Devia, E. Carlini and S. Maggiolo, Journal of Chromatography A **1216**, 6762 (2009).

[115] P. Stefanelli, D. Attard Barbini, S. Girolimetti, A. Santilio and R. Dommarco, J Environ Sci Health B **46**, 341 (2011).

[116] N. Fathima, R. Rao, B. U. Nair, Environ. Engin. Sci. **29**, 363 (2012) doi:10.1089/ees.2010.0445

[117] L. F. Cabeza, M. M. Taylor, G. L. DiMaio, E. M. Brown, W. N. Marmer, R. Carrio and P. J. Celma, J. Cot, Waste Manage, **18**, 211 (1998) doi:10.1016/S0956-053X (98) 00032-4

[118] F. Langmaier, S. Sukop, M. mladek and K. Karel, J. Soc. Leath. Tech. Ch. 83, 187 (1999).

[119] J. Huiyan, L. Junsheng and H. Wei, Waste Manage Res., 34, 399 (2016) doi:10.1177/0734242X16633772

[120] J. Kanagaraj, V. C. Velappan, N. K. Chandra Babu and S. Sadulla, J. Sci. Ind. Res. 65, 541 (2006). doi:10.1002/chin.200649273

[121] M. G. Amparo, J. P. C. Juan and M. V. María, Int. J. Phytoremediat 12, 317 (2010). doi:10.1080/15226510902968118

[122] J. D. Putshaka, A. Yakubu, A. Jauro, H. M. Harami, and M. Kolo, Am. J. chem. Appl. 2, 89 (2015).

[123] S. Srivastava and I. S. Thakur, Bioresour. Technol., 97, 1167 (2006) doi:10.1016/j.biortech.2005.05.012

[124] M. Hamdi, and P. Ibrahim, desalin. Water Treat. 51, 2116 (2012) doi:10.1080/19443994.2013.734589

[125] N. Ibrahim, S. Mustafa, K. Kadriye, E. Latif and D. Mehmit, Anal. Let. 35, 1734 (2007) doi:10.1081/AL-120006679

[126] J. Posta, A. Alimonti, F. Petrucci and S. Caroli, Anal. Chim. Acta. 325, 185 (1996) doi:10.1016/0003-2670(96)00030-X

[127] S. Balasubramanian and V. Pugalenthi, Talanta. 50, 457 (1999) doi:10.1016/S0039-9140(99)00135-6

[128] K. M. Rizwan, M. Naushad and A. Z. Abdullah, Sci. reports, 7, 1707 (2017). https://dx.doi.org/10.1038%2Fs41598-017-01968-x

[129] M. Aznar, E. Canellas and C. Nerín, J. chromatographia A., 1216, 5176 (2009). https://doi.org/10.1016/j.chroma.2009.04.096

[130] M. Ma, X. B. Luo, B. Chen, S. B. Sub and S. Z. Yao, J. Chromatography A., 1103 (1) 170 (2006). https://doi.org/10.1016/j.chroma.2005. 11.061

[131] F. Tateo, and M. J. Bononi, J. Agric. Food Chem., 52(4) 655 (2004) https://pubs.acs.org/action/showCitFormats?doi=10.1021%2Fjf030721s

[132] K. Lacasse and W. Baumann, Textile chemicals: environmental data and facts, (2004) ISBN 3-540 40815-0. Springer Berlin. https://www.springer.com /in/book/9783642623462

[133] B. J. Brüschweiler and C. Merlot, Regul. Toxicol. Pharmacol., 88, 214 (2017) https://doi.org/10.1016/j.yrtph.2017.06.012

[134] T. Robinson, G. McMullan, R. Marchant, and P. Nigam, Bioresource Technology, 77, 247 (2001). https://doi.org/10.1016/S0960-8524(00)00080-8

[135] B. J. Brüschweiler, S. Küng, D. Bürgi, L. Muralt and E. Nyfeler, Regul. Toxicol. Pharmacol., 69(2) 263 (2014). https://doi.org/10.1016/j.yrtph. 2014.04.011

[136] D. Pezo, M. Fedeli, O. Bosetti, C. Nerín, Anal. Chim. acta., 756 49 (2012). https://doi.org/10.1016/j.aca.2012.10.031

[137] D. Rawat, V. Mishra and R. S. Sharma, Chemosphere, 155(1) 591 (2016). https://doi.org/10.1016/j.chemosphere.2016.04.068

[138] BGFA (Forschungsinstitut für Arbeitsmedizin und der Deutschen GesetzlichenUnfallversicherung), 2009. Azofarbmittel und deren Hautg€angigkeit beim Menschen.Literaturstudie. Institut der Ruhr-Universit€at Bochum. BGFA-Report 2 http://www.ipa-

dguv.de/medien/ipa/publikationen/ipa-reporte/09-02-27_bgfareport2_azofarbstoffe.pdf.

[139] T. Platzek, C. Lang, G. Grohmann, U. S. Gi and W. Baltes, Hum. Exp. Toxicol., 18 (9) 552 (1999) https://doi.org/10.1191/096032799678845061

[140] R. L. Stingley, W. Zou, T. M. Heinze, H. Chen, and C. E. Cerniglia, J. Med. Microbial., 59(1) 108 (2010). https://doi.org/10.1099/jmm.0.012617-0

[141] G. Korinth, K. H. Schaller and H. Drexler, Front. Biosci., 5(1) 928 (2013). http://dx.doi.org/10.2741/E672

[142] R. Benigni, L. Passerini, Mutat. Res., 511(3) 191 (2002) https://doi.org/10.1016/S1383-5742(02)00008-X

[143] EC, 2009. Commission Regulation (EC) No 552/2009 of 22 June 2009 Amending Regulation (EC) No 1907/2006 of the European Parliament and of the Council on the Registration, Evaluation, Authorisation and Restriction of Chemicals (REACH) about Annex XVII. http://eur-lex.europa.eu/LexUriServ/ LexUriServ.do? uri¼OJ: L: 2009:164:0007:0031: EN:PDF.

[144] V. Andrisano, R. Gotti, P. Roveri, and V. Cavrini, J. Chromatographia, 44(7-8) 431 (1997) https://doi.org/10.1007/BF02466323

[145] C. R. Noguerol, J. M. Lopez-Vilarino, G. Fernandez-Martinez, L. Barral-Losada and M.V. Gonzalez-Rodriguez, J. Chromatography A., 1179(2) 152 (2008). https://doi.org/10.1016/j.chroma.2007.11.099

[146] M. Shelke, S. K. Sanghi, A. Asthana, S. Lamba and M. Sharma, J. Chromatography A., 1089(1) 52 (2005). https://doi.org/10.1016/ j.chroma.2005.06.029

[147] P. J. Rivero, A. Urrutia, J. Goicoechea and F. J. Arregui, Nanoscale research letters, **10** (1), 501 (2015). doi:10.1186/s11671-015-1195-6

[148] P. Karas, A. Metsoviti, V. Zisis, C. Ehaliotis, M. Omirou, E. S. Papadopoulou, U. Menkissoglou-Spiroudi, S. Manta, D. Komiotis and D. G. Karpouzas, Sci Total Environ **530-531**, 129-139 (2015).

[149] A. K. Salmeia, S. Gaan and G. Malucelli, Polymers **8** (9) 319 (2016).

[150] K. Jarova, A. Koleckarova, Z. Kralova and M. Vavrova, Neuro Endocrinol Lett **37** (Suppl1), 45 (2016).

[151] K. S. Huang, C. H. Yang, S. L. Huang, C. Y. Chen, Y.Y. Lu and Y. S. Lin, International Journal of Molecular Sciences, **17** (9) (2016). doi:10.3390/ijms17091578. (Accession No. 27657043)

[152] V. Giannone, I. Pitino, B. Pecorino, A. Todaro, A. Spina, M. R. Lauro, F. Tomaselli and C. Restuccia, Int J Food Microbiol **235**, 71-76 (2016).

[153] P. A. Karas, S. Makri, E. S. Papadopoulou, C. Ehaliotis, U. Menkissoglu-Spiroudi and D. G. Karpouzas, Ecotoxicol Environ Saf **124**, 447-454 (2016).

[154] S. C. Hammel, K. Hoffman, A. M. Lorenzo, A. Chen, A. L. Phillips, C. M. Butt, J. A. Sosa, T. F. Webster and H. M. Stapleton, Environment International **107**, 181 (2017).

[155] E. Cullen, D. Evans, C. Griffin, P. Burke, R. Mannion, D. Burns, A. Flanagan, A. Kellegher, G. Schoeters, E. Govarts, P. Biot, L. Casteleyn, A. Castaño, M. Kolossa-Gehring, M. Esteban, G. Schwedler, H.M. Koch, J. Angerer, L.E. Knudsen, R. Joas, A. Joas, B. Dumez, O. Sepai, K. Exley and D. Aerts, Int. J. Environ. Res. Public Health **14** (12), 1456 (2017).

[156] L. Cheng, S. Pan, C. Ding, J. He and C. Wang, J. Chromatogr. A **1511**, 85 (2017). doi:10.1016/j. chroma.2017.07.012.

[157] T. T. Yang, J.-P. Guan, R.-C. Tang and G. Chen, Industrial Crops and Products **115**, 16 (2018).

[158] C. Liu, T. Xing, B. Wei and G. Chen, Materials **11** (10) 1842 (2018).

[159] D. Nacci, B. Clark, M. J. La Guardia, K. Miller, D. Champlin, I. Kirby, A. Bertrand and S. Jayaraman, Environmental Toxicology and Chemistry **37** (9), 2350 (2018).

[160] M. E. Street, S. Angelini, S. Bernasconi, E. Burgio, A. Cassio, C. Catellani, F. Cirillo, A. Deodati, E. Fabbrizi, V. Fanos, G. Gargano, E. Grossi, L. Iughetti, P. Lazzeroni, A. Mantovani, L. Migliore, P. Palanza, G. Panzica, A. M. Papini, S. Parmigiani, B. Predieri, C. Sartori, G. Tridenti and S. Amarri, International Journal of Molecular Sciences **19** (6), 1647 (2018).

[161] F. Iadaresta, M. D. Manniello, C. Östman, C. Crescenzi, J. Holmbäck and P. Russo, Environmental Science and Pollution Research **25** (25), 24629 (2018).

[162] H. Yanagisawa, Y. Kudo, K. Nakagawa, H. Miyagawa, F. Maruyama and S. Fujimaki, Molecules **23** (4), 728 (2018).

[163] C. Warembourg, X. Basagaña, C. Seminati, J. de Bont, B. Granum, S. Lyon-Caen, C.B. Manzano-Salgado, I. Pin, A.K. Sakhi, V. Siroux, R. Slama, J. Urquiza, M. Vrijheid, C. Thomsen and M. Casas, Int. J. Hyg. Environ. Health, **222,** 446 (2018).

[164] R. S. Strakovsky and S.L. Schantz, Environ. Epigenet. **4** (3) (2018). doi:10.1093/eep/dvy022.

[165] D.L. McBride, J. Pediatri. Nurs. **45,** 76 (2018).

[166] N. Bandow, S. Gartiser, O. Ilvonen and U.J.E.S.E. Schoknecht, Environ. Sci. Eur. **30,** 14 (2018).

[167] C. Ursino, R. Castro-Muñoz, E. Drioli, L. Gzara, M.H. Albeirutty and A. Figoli, Membranes **8** (2), 18 (2018).

[168] K. Minjeong, J. Ji Seong, K. Hyunji, H. Seungwoo, P. Il-Hyun, L. Byung-Chul, Y. Sung Il, J.S. Ha, N.K. Taek and L. Kyung-Min, Biomol. Ther. (Seoul) **26,** 512 (2018). doi:10.4062/ biomolther.2018.076.

[169] D. Höfer, Dermatology research and practice, 2018, 7043438 (2018). doi:10.1155/2018/7043438. (Accession No. 30581460)

[170] H. Kuwahara, J. Ninomiya and H. Morita, Biocontrol Science **23** (3), 85-96 (2018).

[171] N. Hosoya, K. Motomura, E. Tagawa, C. Ogihara, M. Nagano and H. Hosoya, *Effects of the fungicide ortho-phenylphenol (OPP) on the early development of sea urchin eggs*. Mar Environ Res **143**, 24-29 (2018).

[172] M. Wannick, J. C. Assmann, J. F. Vielhauer, S. Offermanns, D. Zillikens, C. D. Sadik and M. Schwaninger, Front Immunol **9** 1890 (2018).

[173] P. Perini, F. Rinaldi, M. Puthenparampil, M. Marcon, F. Perini and P. Gallo, Mult Scler Relat Disord **26**, 68 (2018).

[174] J. L. Andersen, B. Gesser, E. D. Funder, C. J. F. Nielsen, H. Gotfred-Rasmussen, M. K. Rasmussen, R. Toth, K. V. Gothelf, J. S. C. Arthur, L. Iversen and P. Nissen, Nature Communications **9**, 4344 (2018).

[175] A. A. Ahmed, A. A. E. Ahmed, E. M. El Morsy and S. Nofal, J Pharm Pharmacol **70**, 1521 (2018).

[176] D. Ontaneda, J. Nicholas, M. Carraro, J. Zhou, Q. Hou, J. Babb, K. Riester, J. P. Mendoza, T. Livingston and M. Jhaveri, Mult Scler Relat Disord **27**, 101 (2018).

[177] C. Wang, J. Sep. Sci. **42**, 1248–1256-2019 v.1242 no.1246 (2019). doi:10.1002/jssc.201801171.

[178] Y. Zhang, H. Su, M. Ya, J. Li, S.-H. Ho, L. Zhao, K. Jian, R. J. Letcher and G. Su, Science of The Total Environment **693**, 133654 (2019).

[179] M. Brits, S. H. Brandsma, E. R. Rohwer, J. De Vos, J. M. Weiss and J. de Boer, Environmental Pollution **253**, 120 (2019).

[180] C. F. Estill, J. Slone, A. C. Mayer, K. Phillips, J. Lu, I. C. Chen, A. Christianson, R. Streicher, M. J. L. Guardia, N. Jayatilaka, M. Ospina and A. M. Calafat, Journal of Occupational and Environmental Hygiene **16** (7), 477 (2019).

[181] Y. J. Hu, L. J. Bao, C. L. Huang, S.-M. Li and E. Y. Zeng, Environmental Pollution **252**, 1902 (2019).

[182] G. Rivière, J. Jean, S. Gorecki, M. Hulin, M. Kolf-Clauw, C. Feidt, N. Picard-Hagen, P. Vasseur, B. Le Bizec and V. Sirot, Food and Chemical Toxicology **131**, 110561 (2019).

[183] M. J. Piazza and A. A. Urbanetz, in *JBRA Assist Reprod* **23**, 154 (2019)

[184] Y. Liang, X. Liu and M. R. Allen, Environmental Science & Technology **52** (10), 5821 (2018).

[185] A. M. Al-Salem, Q. Saquib, M. A. Siddiqui, J. Ahmad, R. Wahab and A. A. Al-Khedhairy, Chemosphere **237**, 124519 (2019).

[186] H. Kang, J. Lee, J. P. Lee and K. Choi, Environment International **131**, 105034 (2019).

[187] S. Yuan, Y. Han, M. Ma, K. Rao, Z. Wang, R. Yang, Y. Liu and X. Zhou, Environmental Pollution **250**, 58 (2019).

[188] Y.Q. Ren, G.H. Wang, J.J. Li, C. Wu, C. Cao, J. Li, J.Y. Wang, S.S. Ge, Y.N. Xie, X.R. Li, F. Meng and H. Li, Chemosphere **215**, 413 (2019). doi:10.1016/j.chemosphere.2018.10.064.

[189] R.M. Martinez, R. Hauser, L. Liang, A. Mansur, M. Adir, L. Dioni, C. Racowsky, V. Bollati, A.A. Baccarelli and R. Machtinger, Environ. Int. **123,** 20 (2019). doi:10.1016/j.envint.2018.11.043.

[190] R.M. Shaffer, K.K. Ferguson, L. Sheppard, T. James-Todd, S. Butts, S. Chandrasekaran, S.H. Swan, E.S. Barrett, R. Nguyen, N. Bush, T.F. McElrath and S. Sathyanarayana, Environ. Int. **123,** 588 (2019). doi:10.1016/j.envint.2018.12.021.

[191] M. Long, M. Ghisari, L. Kjeldsen, M. Wielsøe, B. Nørgaard-Pedersen, E.L. Mortensen, M.W. Abdallah and E.C.J.M.A. Bonefeld-Jørgensen, Mol. Autism **10** (1), 1 (2019).

[192] D. P. Bockmühl, J. Schages and L. Rehberg, Microbial cell (Graz, Austria), **6** (7), 299 (2019). doi:10.15698/mic2019.07.682. (Accession No. 31294042)

[193] A. de la Torre, I. Navarro, P. Sanz and M. de los Ángeles Martínez, Journal of Hazardous Materials **382**, 121009 (2020).

[194] N. Wemken, D. S. Drage, C. Cellarius, K. Cleere, J. J. Morrison, S. Daly, M. A.-E. Abdallah, C. Tlustos, S. Harrad and M. A. Coggins, Environmental Research **180**, 108805 (2020).

PUBLICATIONS

Development of a new gas chromatographic-mass spectrometry method for the simultaneous analysis of all regulated phthalates in consumer goods. International Journal of Environmental Analytical Chemistry. https://doi.org/10.1080/03067319.2019.1651850

Development of an Efficient Analytical Method for the Extraction and Analysis of Biocide Contents from the Textile Test Specimens on LC-DAD. Journal of Analytical Methods in Chemistry https://doi.org/10.1155/2020/3047961

Synthesis, In vitro, and In silico Investigations of 4-(1H-Indol-3-yl)-N'-[(E/Z)-(phenyl-substituted)methylidene] as Effective Inhibitors of α-Glucosidase. Russian Journal of Bio-organic chemistry. http://dx.doi.org/10.1134/S1068162024050157

Exploring the Mechanisms of Indole-Oxadiazole Benzamide Hybrids as Tyrosinase Inhibitors: Insights from Lineweaver-Burk Plot Analysis and Computational Studies. Russian Journal of Bio-organic chemistry. https://doi.org/10.1134/S106816202406013X

Electrochemically Synthesized Ni-Mo Alloy via Hydrogen Evolution Reaction: Characterization and Performance Analysis https://papers.ssrn.com/abstract=5155285

Efficiency Enhancement of Dyes Sensitized Solar Cells through Co-Sensitization: A Comprehensive Review https://papers.ssrn.com/abstract=5155465

Extraction and Characterization of Fixed Oil from Spinacia oleracea Seeds: Physicochemical Properties and Bioactive Compounds https://papers.ssrn.com/abstract=5166118

Development of an efficient Gas Chromatography-Mass Spectrometry (GC-MS) method for the estimation of extractable ortho-phenylphenol (OPP) contents in textile and leather test specimens https://papers.ssrn.com/abstract=5173914

A novel validated method for extraction and estimation of Dimethyl fumarate from various textile matrices on GCMS https://papers.ssrn.com/abstract=5173851

Trace Metals Detection in Drinking Water by analytical techniques, comparison of brands and Implications for Public Health and Environmental Safety https://papers.ssrn.com/abstract=5170610

Detection and Quantification of Hydroquinone in Whitening Creams Using UV-Visible Spectrophotometry: Health Implications and Regulatory Concerns https://papers.ssrn.com/abstract=5170525

Heavy Metal Contamination in Cosmetics: An Analysis of Lead and Mercury Levels in Commonly Used Products and Associated Health Risks https://papers.ssrn.com/abstract=5169039

Detection and Quantification of Steroid Adulteration in Herbal Medicines Using TLC and UV-Spectroscopy https://papers.ssrn.com/abstract=5167828

Impact of Poultry Growth Stages on Toxic Compounds Formation During Charcoal Grilling https://papers.ssrn.com/abstract=5160820

Study of ZnO NPs and modified MEG (Mono-ethylene glycol)/ ZnO-NPs in drug loading efficiency and dye degradation applications https://papers.ssrn.com/abstract=5159667

Highly Sensitive Method for the Quantitative Analysis of Polycyclic Aromatic Hydrocarbons (PAHs) in Consumer Articles Using GCMS

Synthesis, in vitro and in silico studies of S-alkylated 5-(4-methoxyphenyl)-4phenyl-4H1,2,4-triazole-3-thiols as cholinesterase inhibitors.

IDENTIFICATION OF VOLATILE CONSTITUENTS OF ETHYL ACETATE FRACTION OF CHENOPODIUM QUINOA ROOTS EXTRACT BY GC-MS

Pesticidal constituents in n-hexane inflorescence extract of Chenopodium quinoa

Evaluation of Elemental Concentrations of PM10 Aerosols from an Industrial Area in Mega City, Lahore (Pakistan): Implications for Air Quality and Human Health. International Journal of Environmental Analytical Chemistry (Submitted).

Optimized and validated LCMS/MS method for simultaneous analysis of all regulated perfluorinated organic compounds (PFOCs) from consumer articles. Journal of Hazardous Materials (Submitted)

Efficiency Enhancement of Dyes Sensitized Solar Cells through Co-Sensitization: A Comprehensive Review" Journal of Solar Energy (Submitted)

Electrochemically Synthesized Ni-Mo Alloy via Hydrogen Evolution Reaction: Characterization and Performance Analysis, Journal of the Chinese Institute of Engineers (Submitted)

Jahangir, M., Khan, U. **Estimation of Bis(2,3-dibromopropyl) phosphate (BDBPP) and Tris(2,3-dibromopropyl) phosphate (TDBPP) by a newly established method on HPLC/MS.**

Jahangir, M., Khan, U. **Extraction of Tris-(aziridinyl) phosphine oxide (TEPA) by sonication using novel method on GC-NPD/MS in textile test specimens.**

Jahangir, M., Khan, U. **Examination of phosphate-based flame retardants from the textile, leather and plastic material by novel method on GC/MS.**

Jahangir, M., Khan, U. **Extraction and estimation of Bis-(2,3-dibromopropyl ether) of Tetrabromobisphenol A (Bis-TBBPA) from the textile and plastic test specimens by GCMS.**

Jahangir, M., Khan, U. **Examination of Dimethyl fumarate in leather, PU leather, textile, rubber and silica gel pellet using a novel method on GCMS technique.**

Jahangir, M., Khan, U. **Extraction and estimation of the contents of the extractable ortho-phenylphenol (OPP), its salts and esters in the textile and leather test specimens by alkaline digestion using newly established method on GCMS.**

Jahangir, M., Khan, U. **Estimation of glycols in textile and leather test specimens using novel method on GCMS.**

Jahangir, M., Khan, U. **Simultaneous determination of glyoxal and short chain aldehydes (acrolein, crotonaldehyde, glutaraldehyde, propanol, butanal, pentanal and hexanal) contents using newly established method HPLC-PDA method.**

Jahangir, M., Khan, U. **Simultaneous determination of N, N-Dimethylformamide, N-Methylformamide, acetophenone, 2-phenyl-2-propanol and formamide in polyethylene (PE) polystyrene (PS), polyurethane (PU) and ethylene vinyl acetate (EVA) by a novel method on GCMS.**

Jahangir, M., Khan, U. **Examination of cationic surfactants dehydrogenated tallowdimethyl ammoniumchloride (DHTDMAC), Distearyldimethyl ammoniumchloride (DSDMAC) and Ditallowdimethyl ammoniumchloride (DTDMAC) by a novel method on HPLC/MS.**

ANNEXURE 1

Mass spectrums and peaks profiles of flame retardants

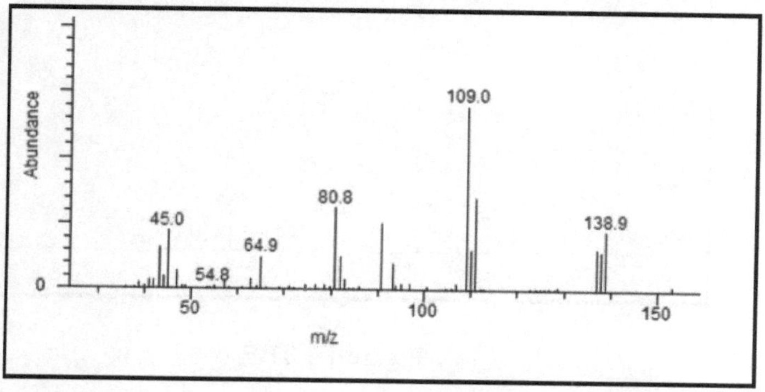

Figure 1.1 TMP

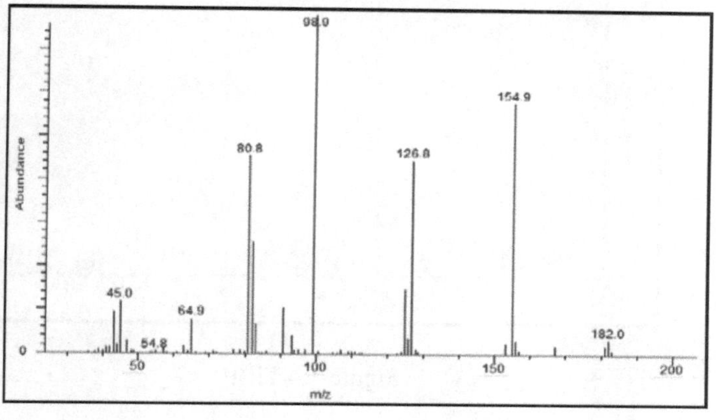

Figure 1.2 TEP

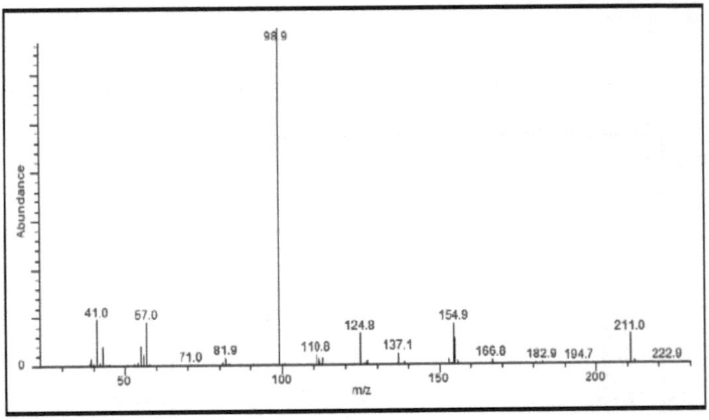

Figure 1.3 TBP

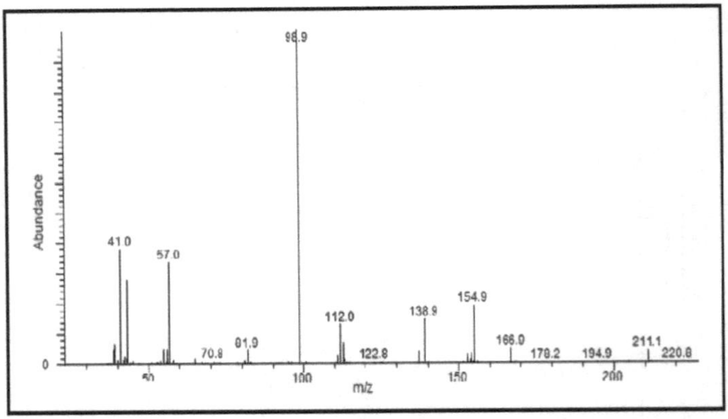

Figure 1.4 TiBP

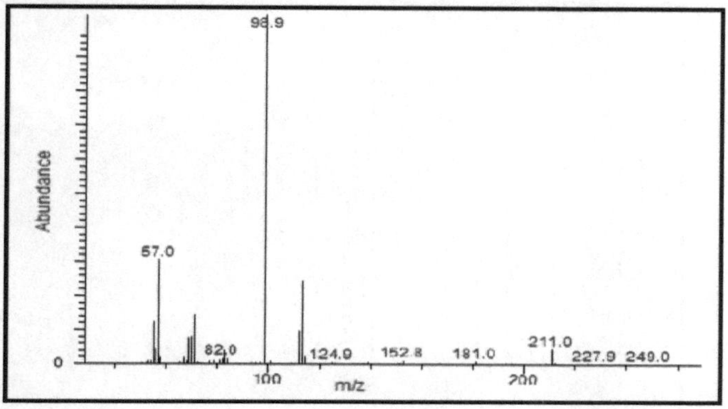

Figure 1.5 TEHP

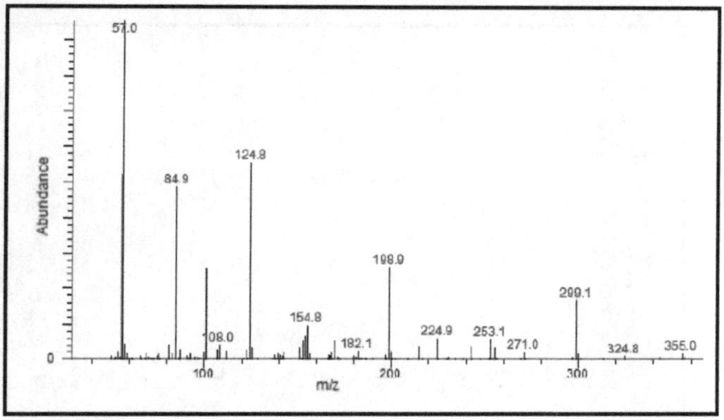

Figure 1.6 TBEP

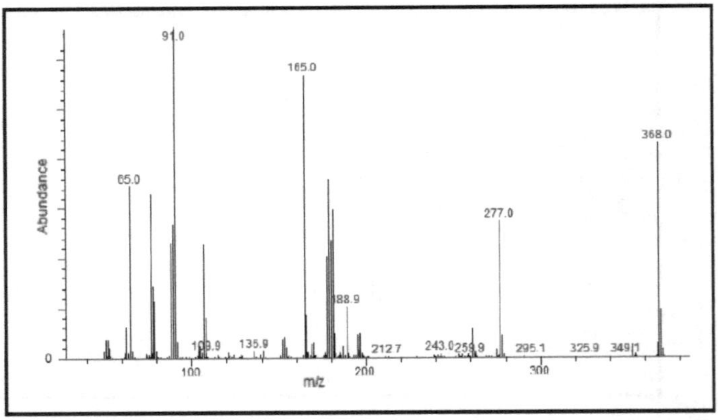

Figure 1.7 TOCP

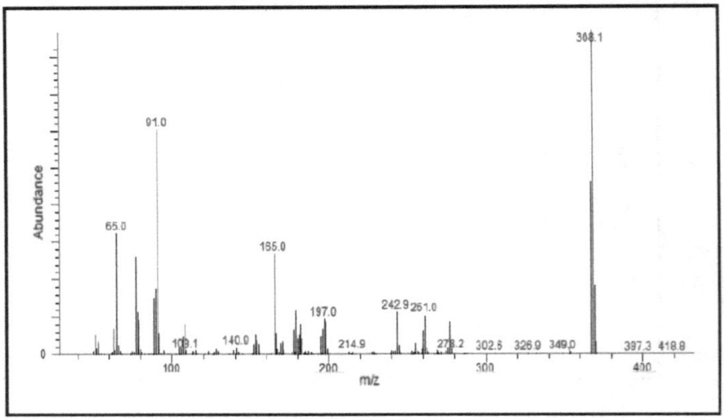

Figure 1.8 TMCP

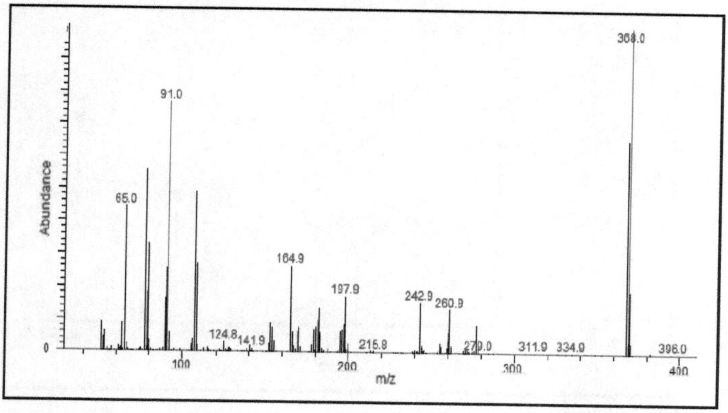

Figure 1.9 TPCP

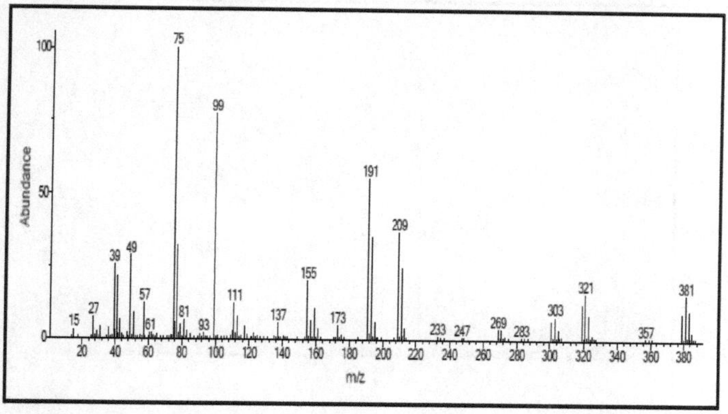

Figure 1.10 TDCPP

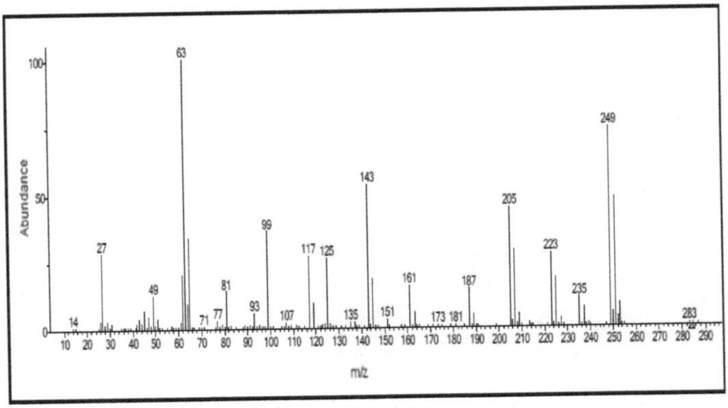

Figure 1.11 TCEP

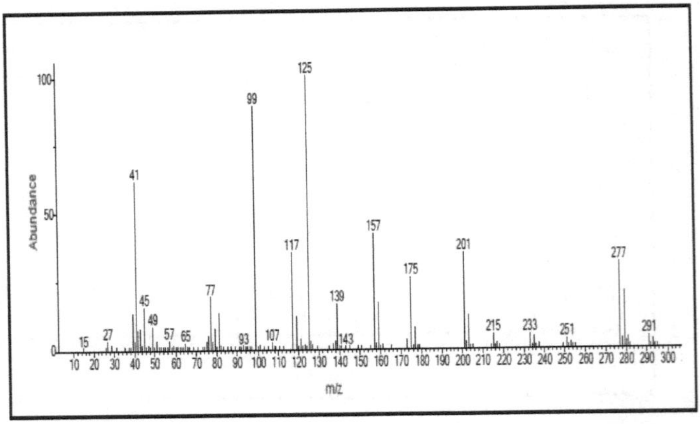

Figure 1.12 TCPP

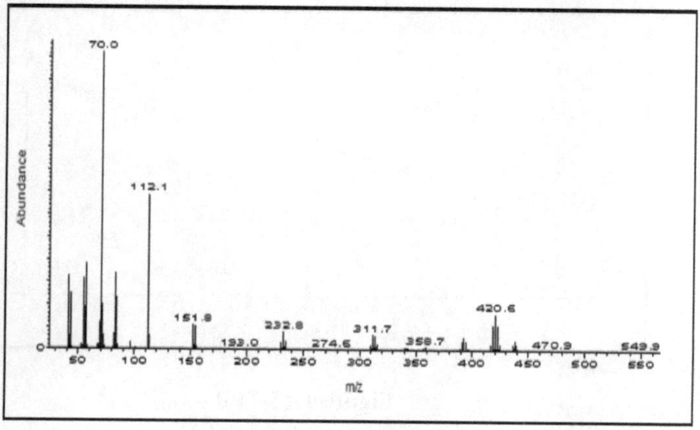

Figure 1.13 TBB

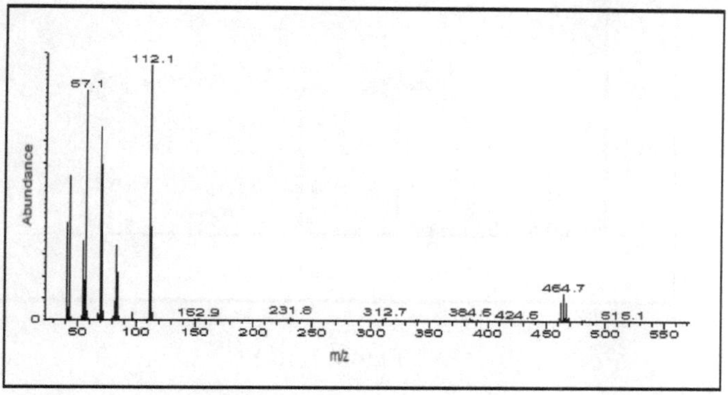

Figure 1.14 TBPH

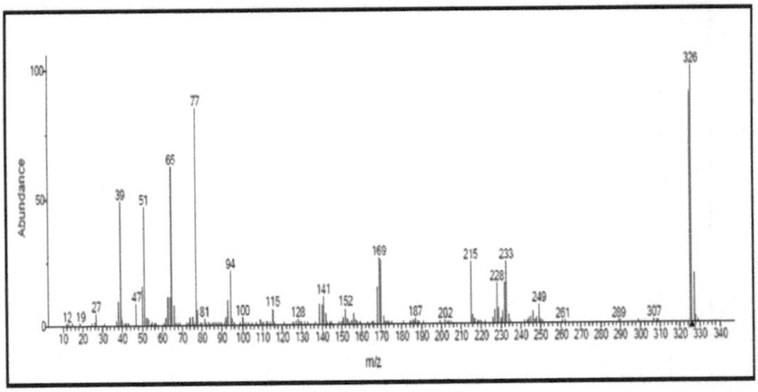

Figure 1.15 TPP

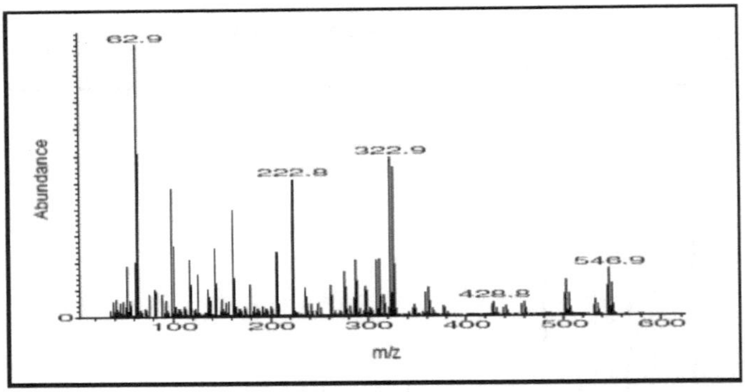

Figure 1.16 V6

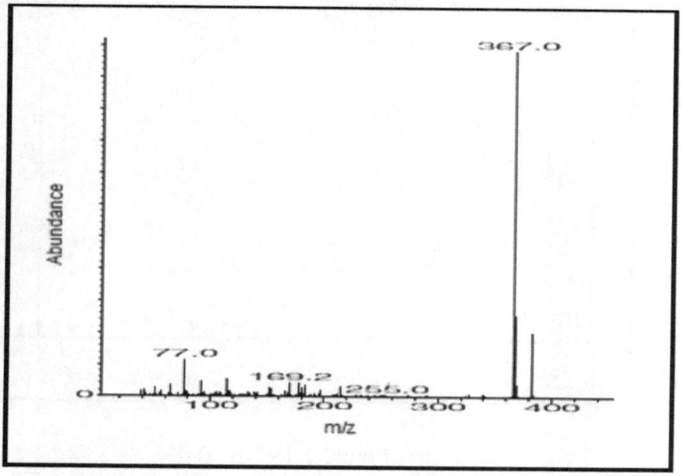

Figure 1.17 MDPP

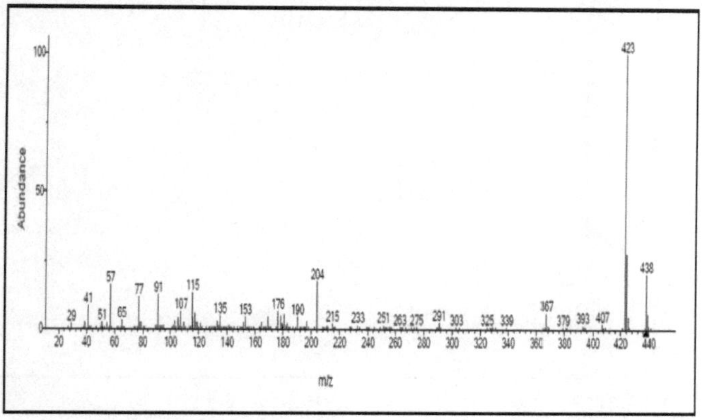

Figure 1.18 TMP

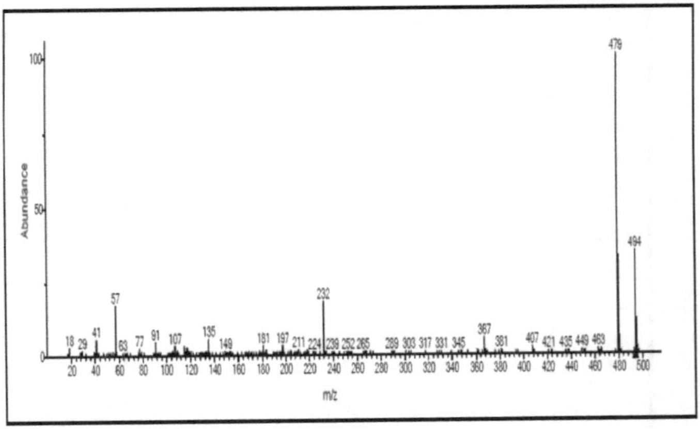

Figure 1.19 TBPP

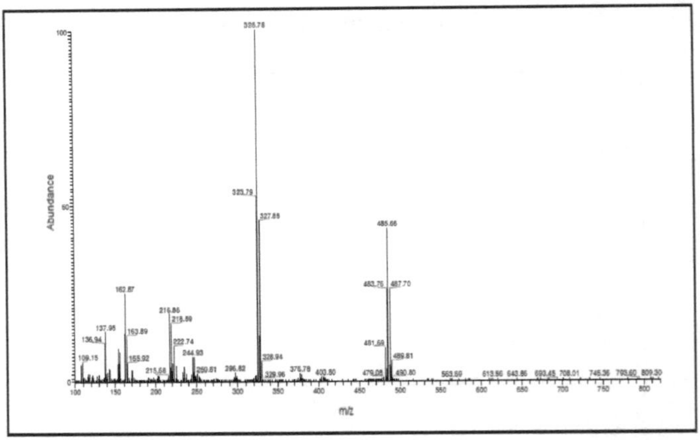

Figure 1.20 Tetra-BDE

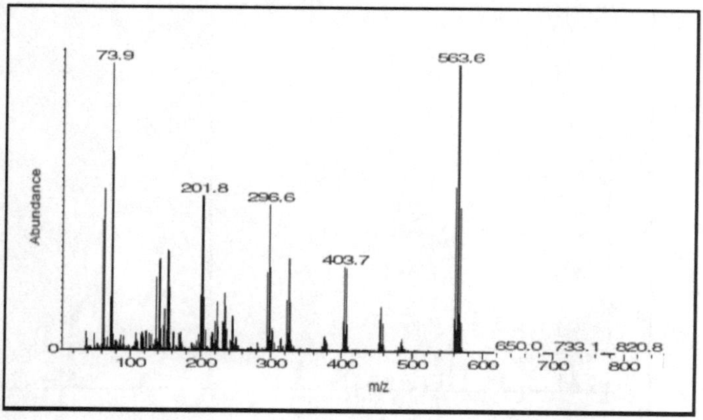

Figure 1.21 Penta-BDE

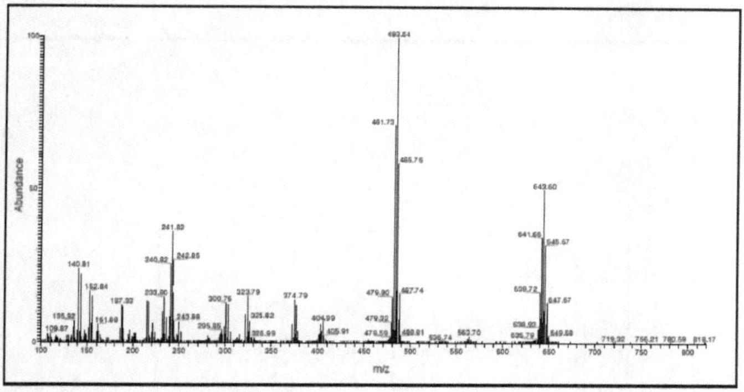

Figure 1.22 Hexa-BDE

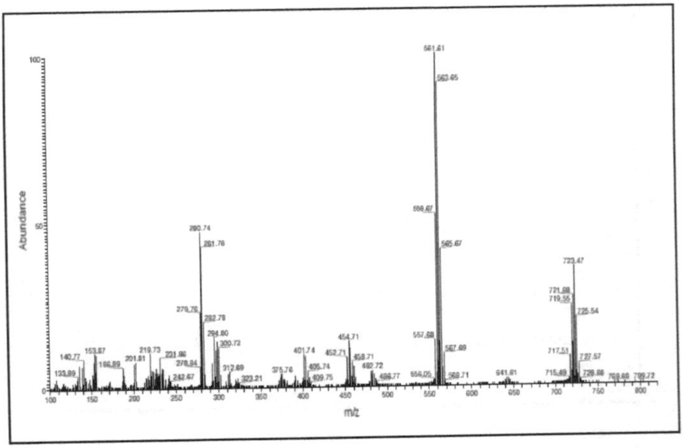

Figure 1.23 Hepta-BDE

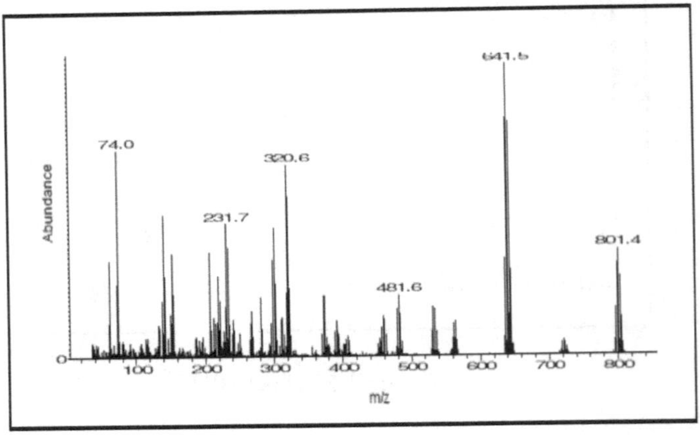

Figure 1.24 Octa-BDE

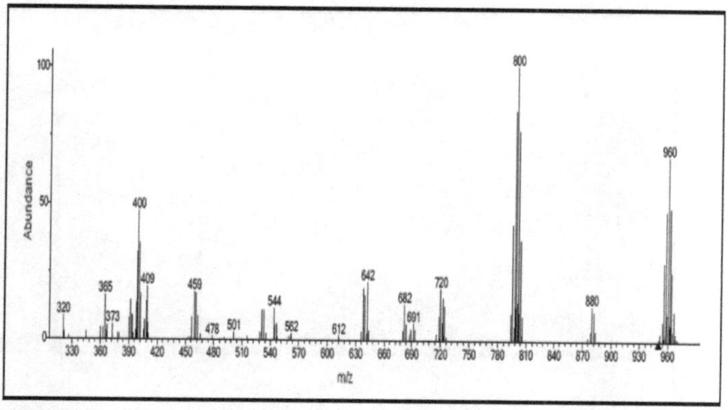

Figure 1.25 Deca-BDE

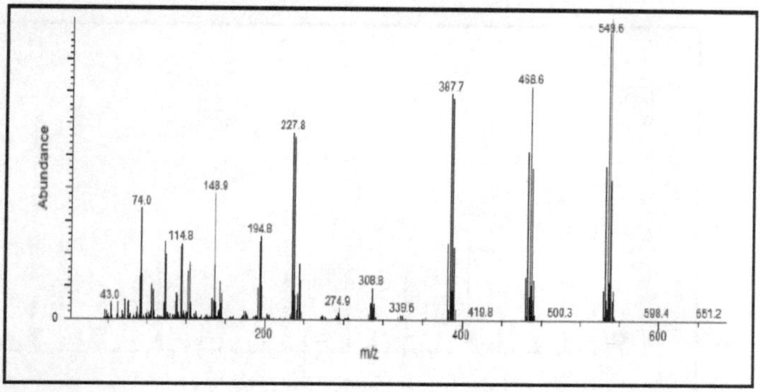

Figure 1.26 Penta-BB

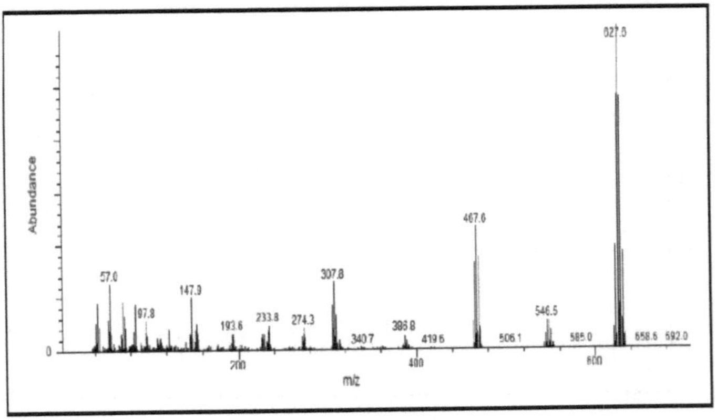

Figure 1.27 Hexa-BB

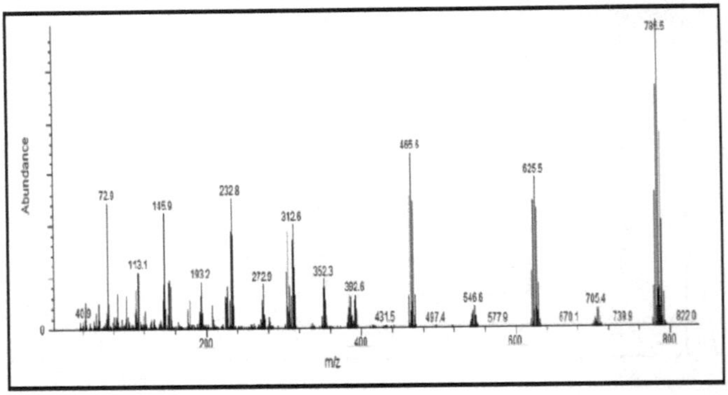

Figure 1.28 Octa-BB

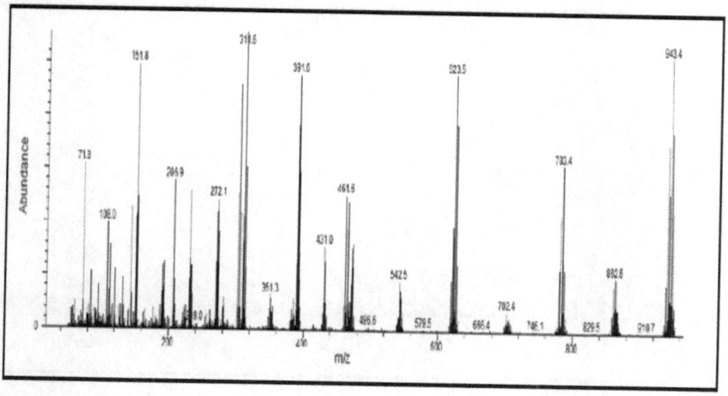

Figure 1.29 Deca-BB

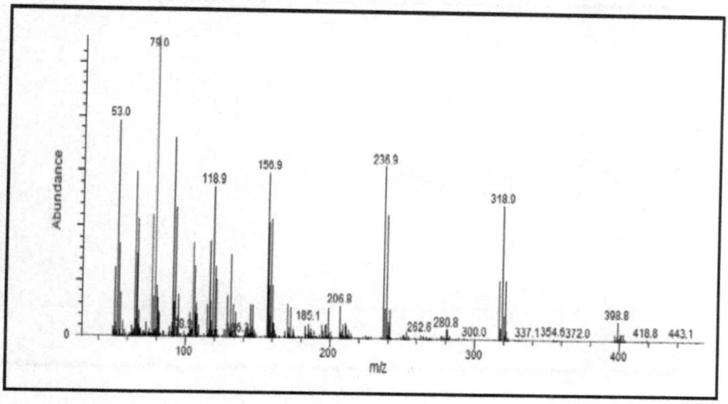

Figure 1.30 HBCDD

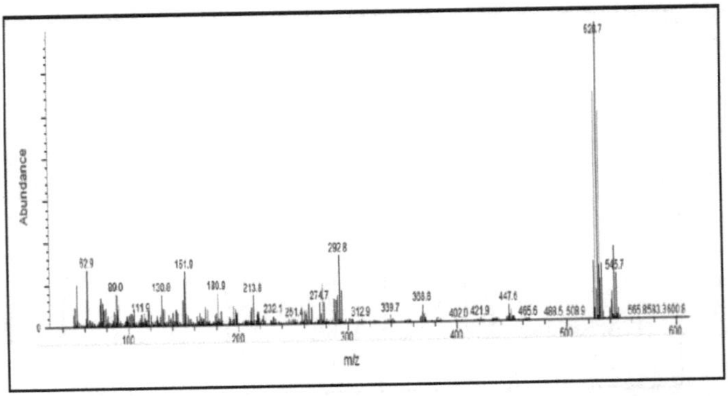

Figure 1.31 TBBPA

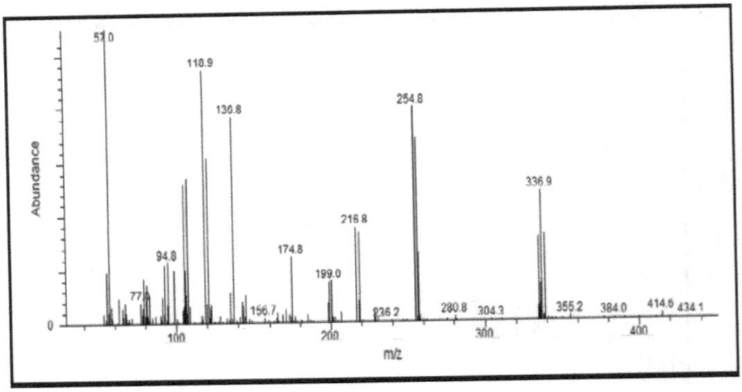

Figure 1.32 BDBPP

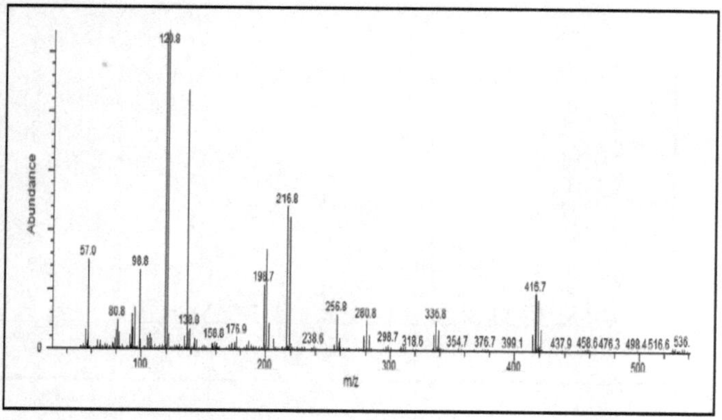

Figure 1.33 TDBPP

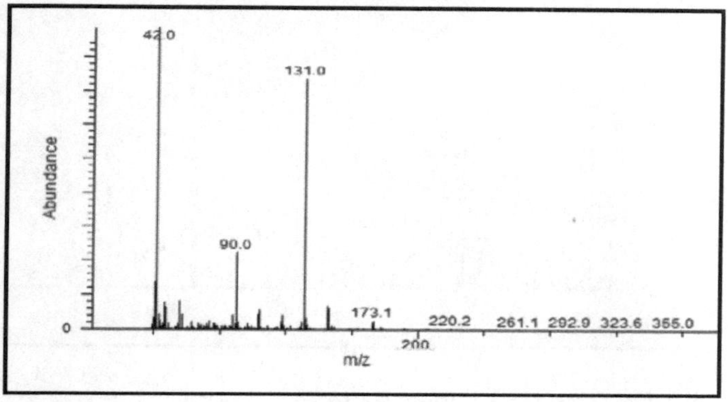

Figure 1.34 TEPA

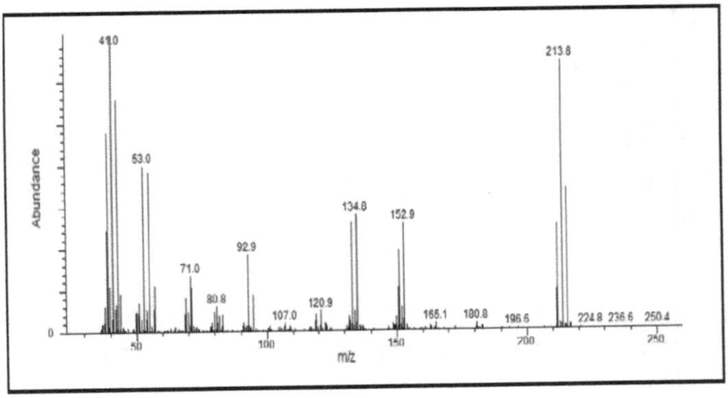

Figure 1.35 BBMP

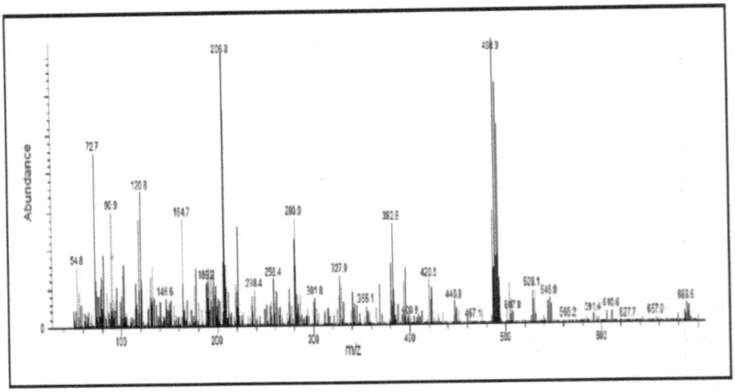

Figure 1.36 Bis-TBBPA

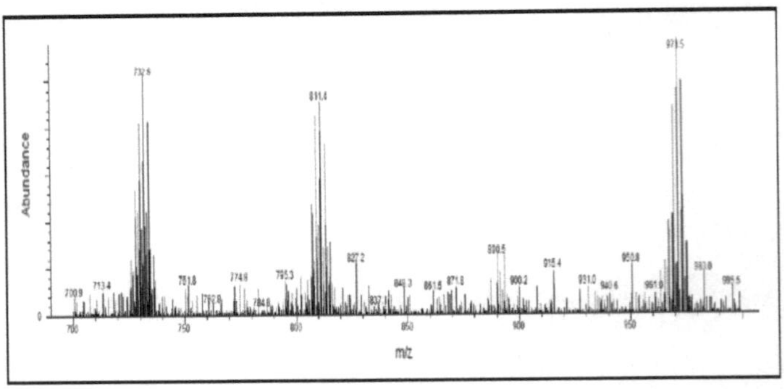

Figure 1.37 DBDPE

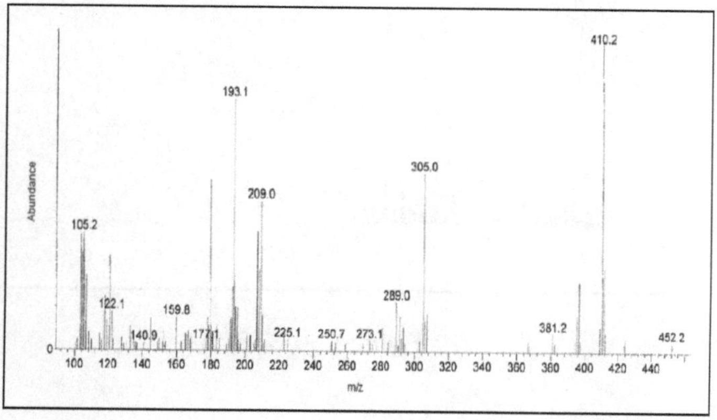

Figure 1.38 TXP

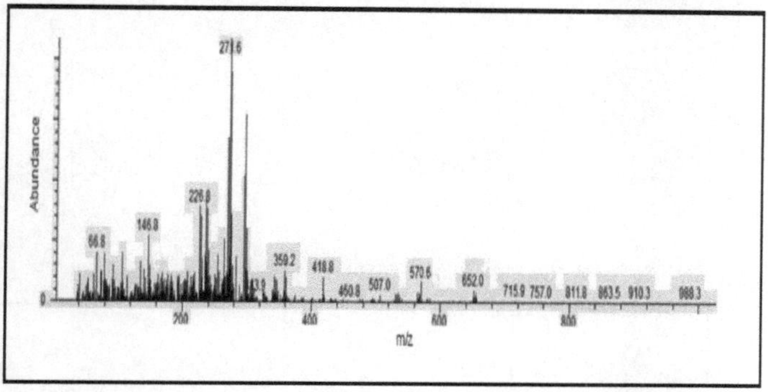

Figure 1.39 Dechlorane Plus

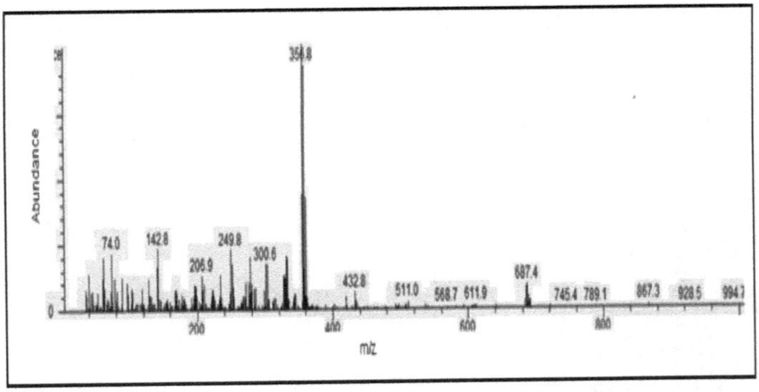

Figure 1.40 BTBPE

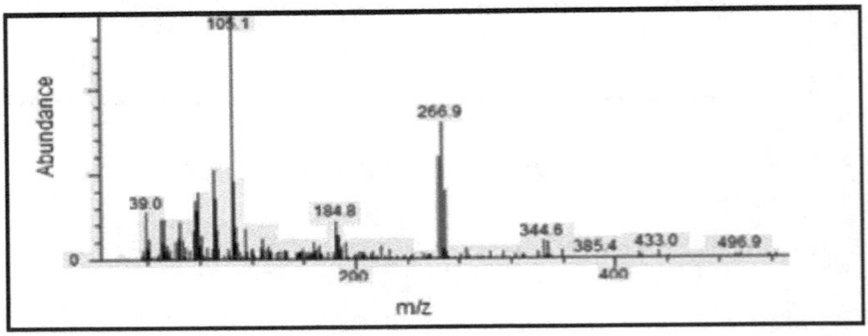

Figure 1.41 TBECH

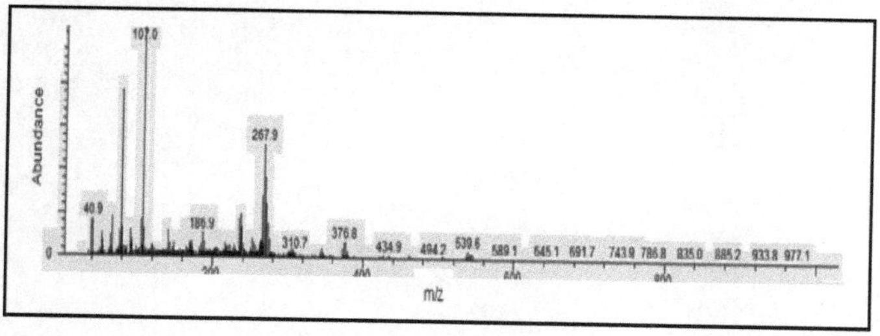

Figure 1.42 DBHC-TCTD

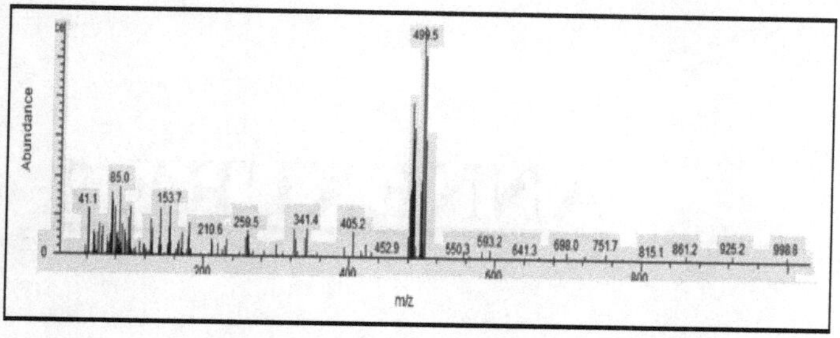

Figure 1.43 PBEB

ANNEXURE 2

Mass spectrums and peaks profiles of phthalates (Data from instrument)

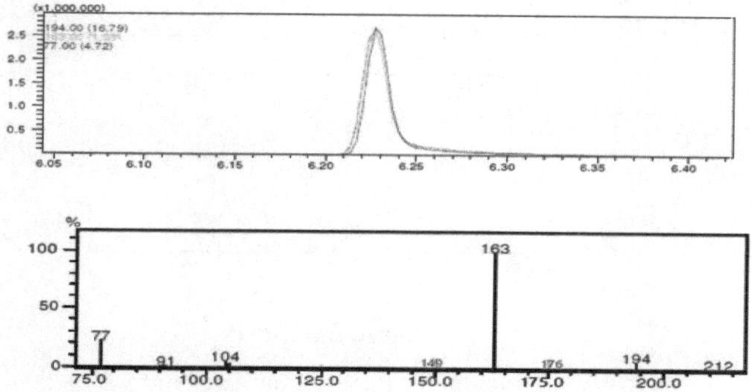

Figure 2.1 DMP

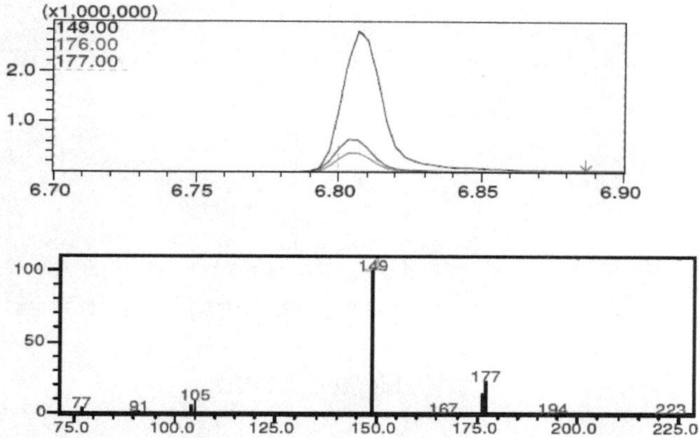

Figure 2.2 DEP

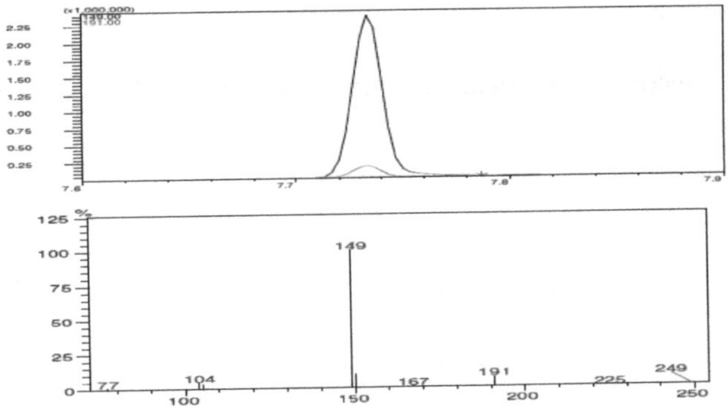

Figure 2.3 DPrP

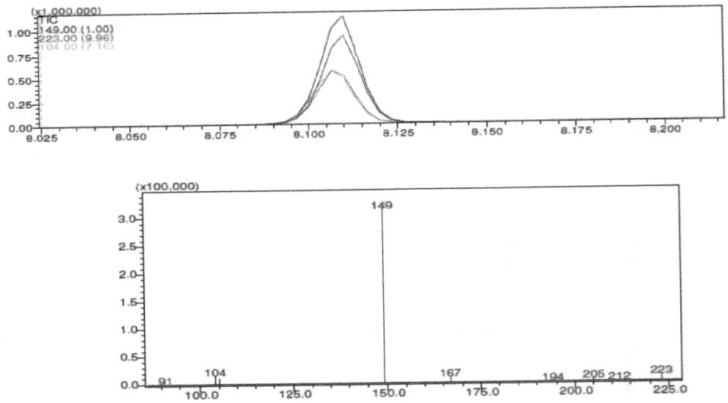

Figure 2.4 DIBP

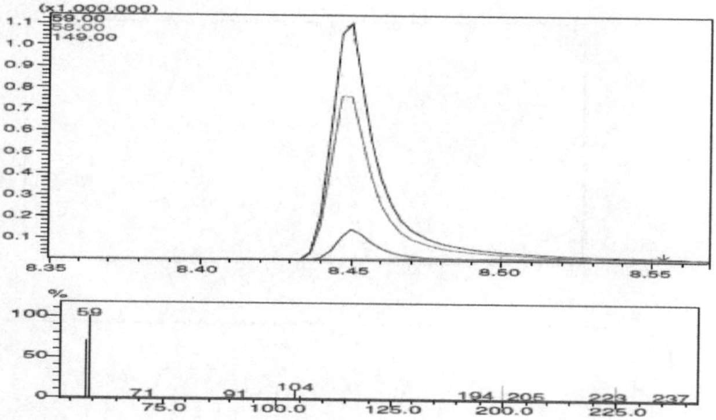

Figure 2.5 DMEP

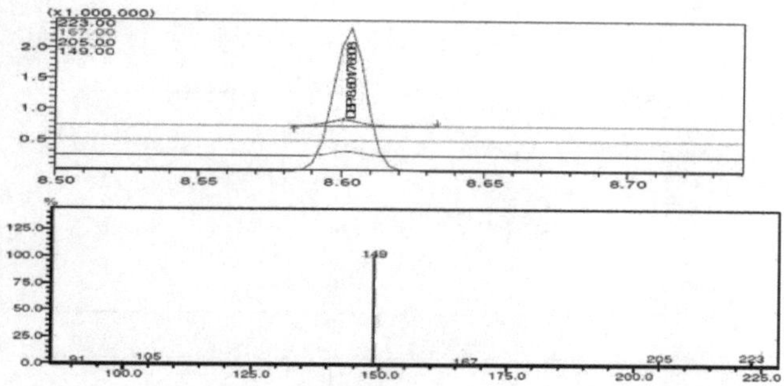

Figure 2.6 DBP

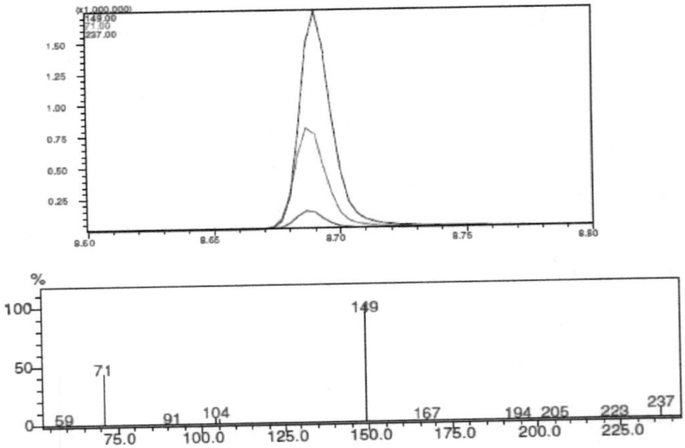

Figure 2.7 DIPP

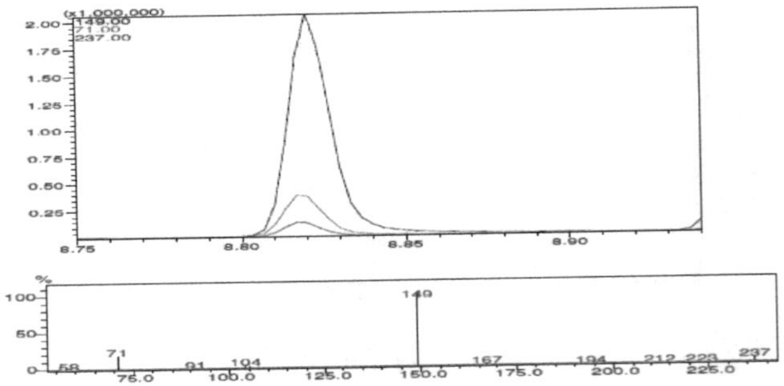

Figure 2.8 nPIPP

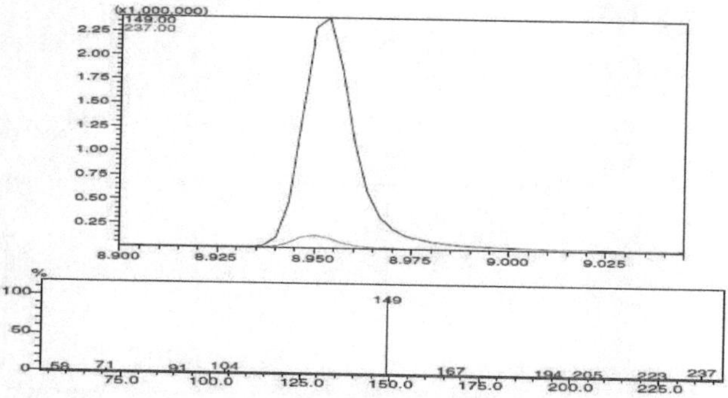

Figure 2.9 DnPP

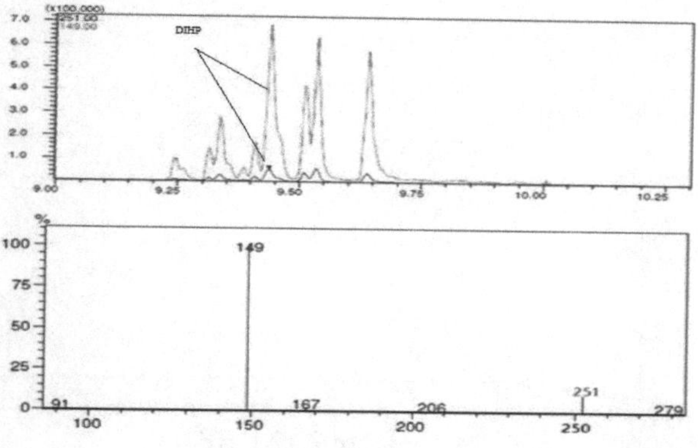

Figure 2.10 DIHP

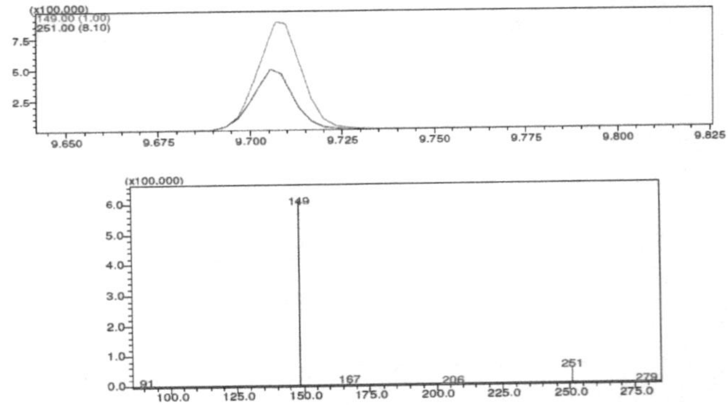

Figure 2.11 DnHP

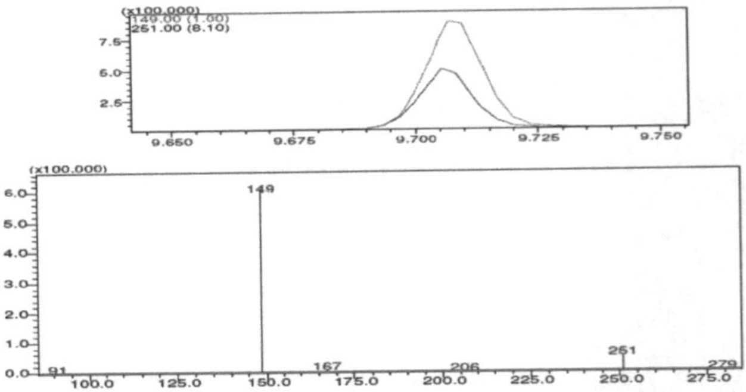

Figure 2.12 DHP

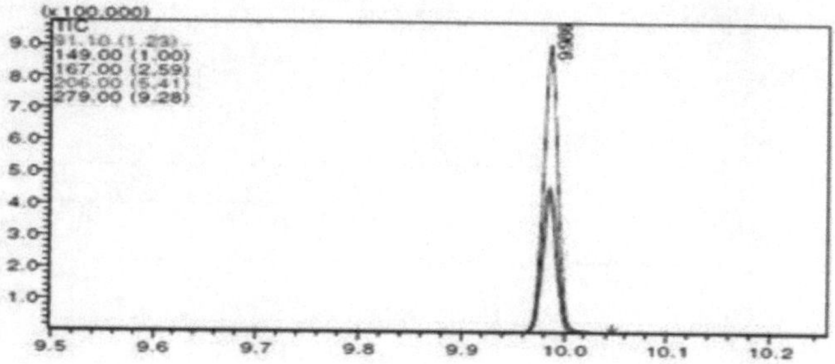

Figure 2.13 BBP

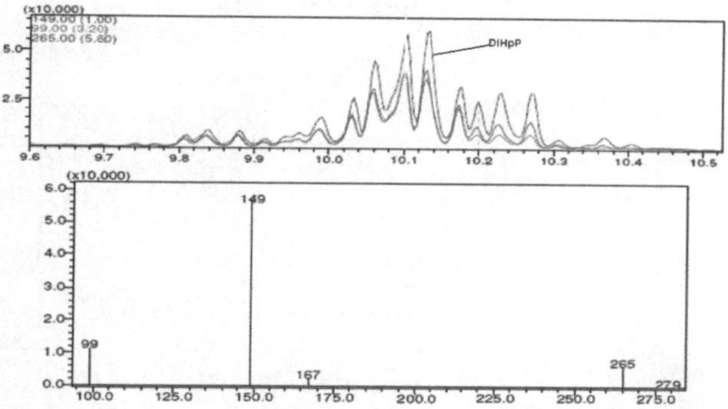

Figure 2.14 DIHpP

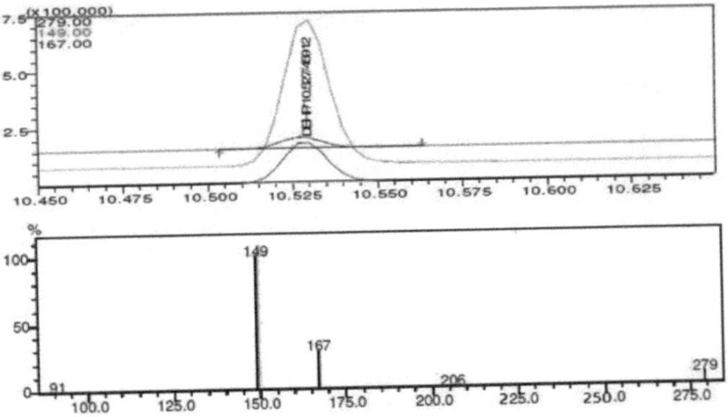

Figure 2.15 DEHP

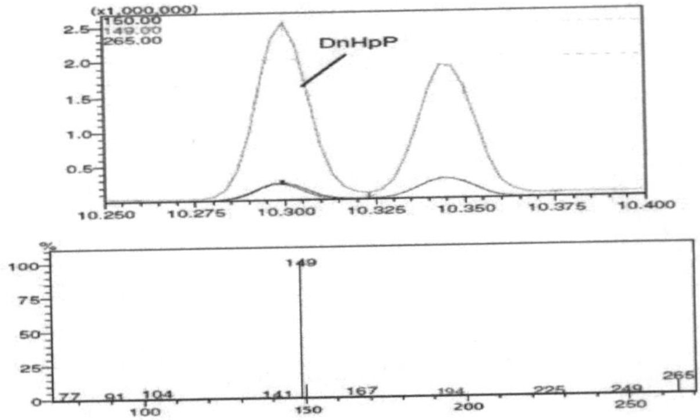

Figure 2.16 DnHpP

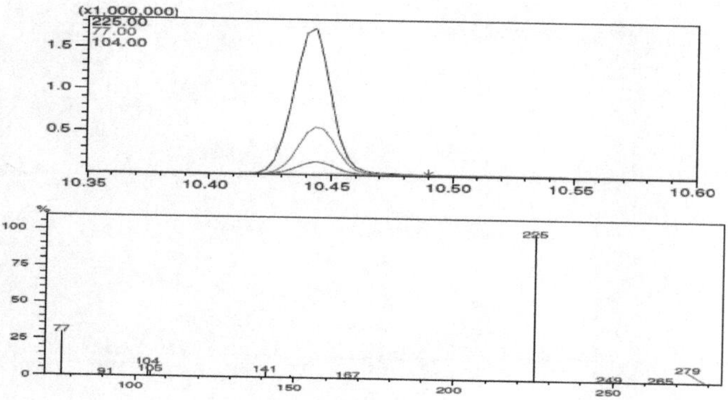

Figure 2.17 DPhP

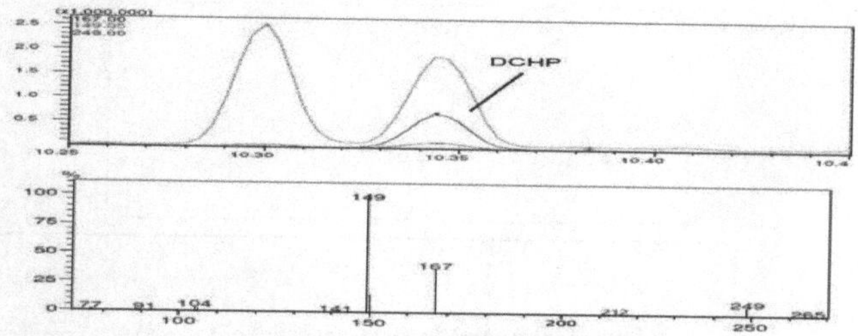

Figure 2.18 DCHP

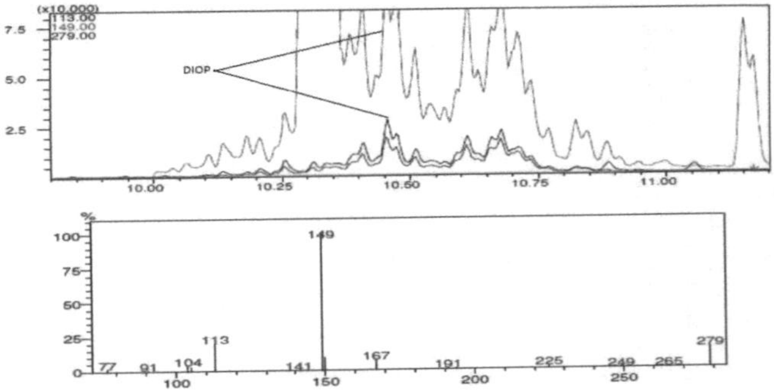

Figure 2.19 DIOP

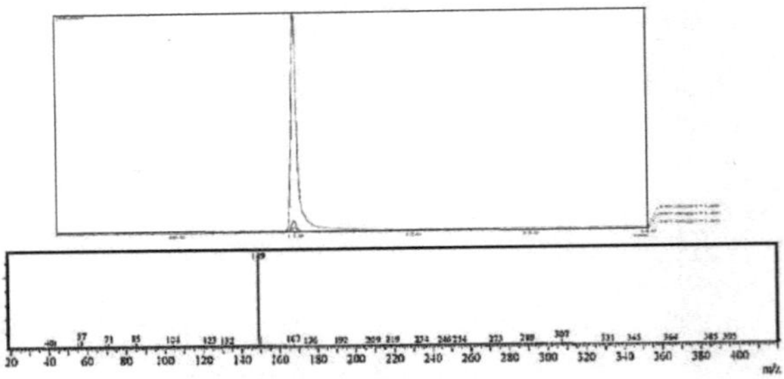

Figure 2.20 DDP

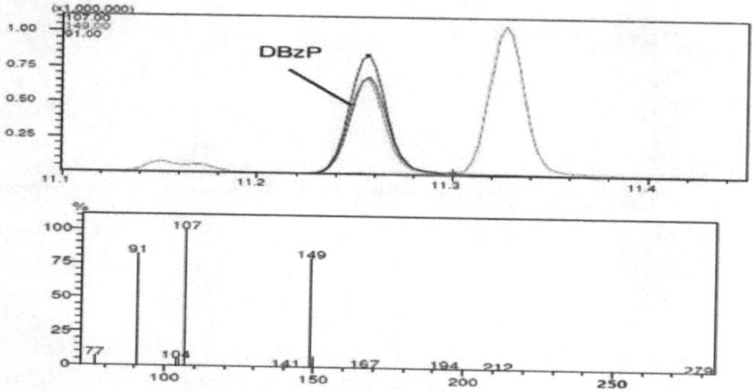

Figure 2.21 DBzP

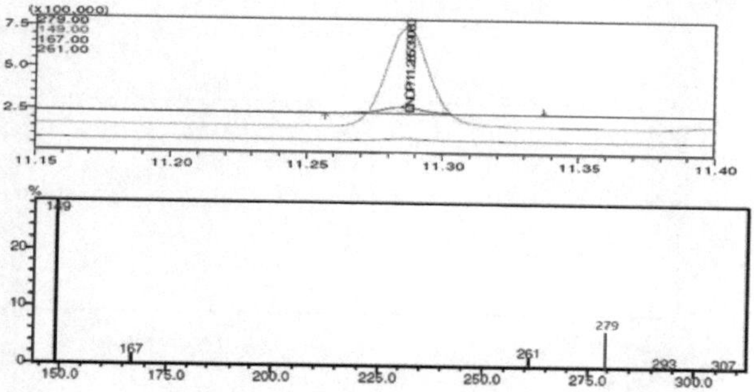

Figure 2.22 DnOP

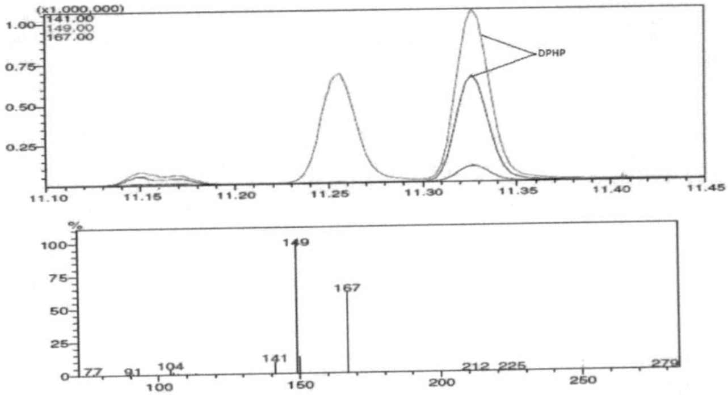

Figure 2.23 DPHP

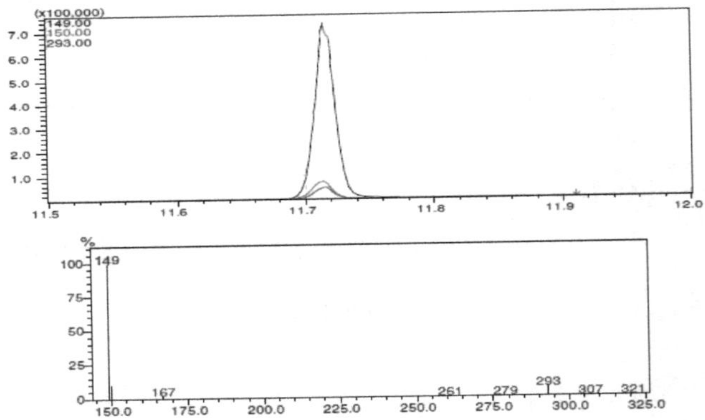

Figure 2.24 DNP

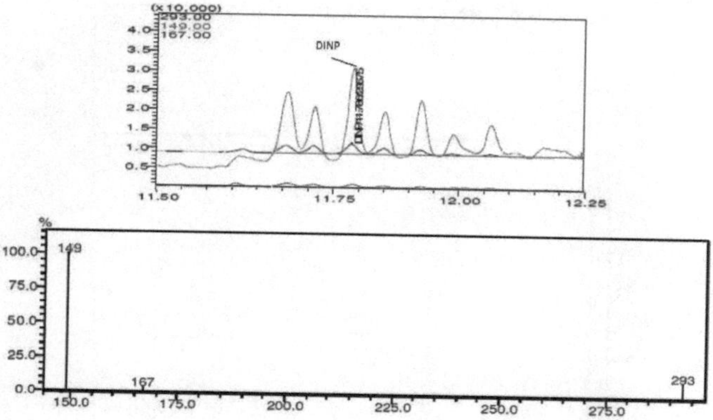

Figure 2.25 DINP

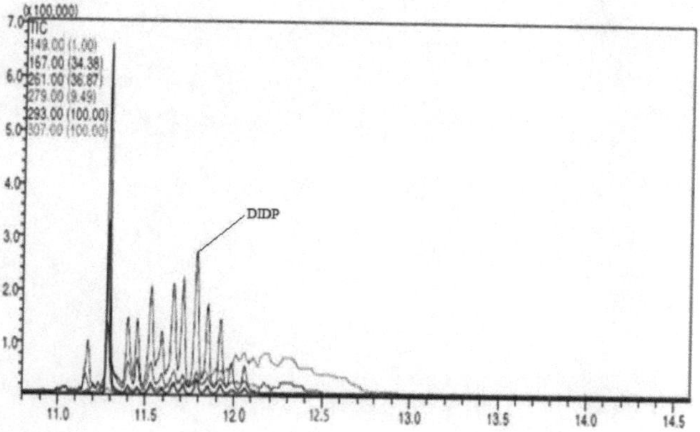

Figure 2.26 DIDP

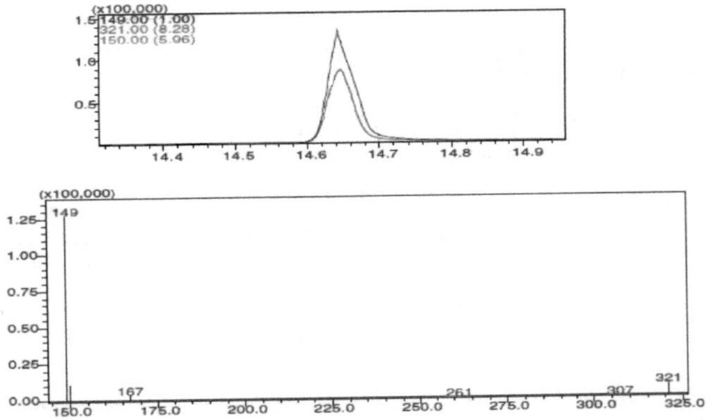

Figure 2.27 DUDP

FSC
www.fsc.org
MIX
Papier | Fördert gute Waldnutzung
FSC® C083411

Kolibri 360 GmbH
Ferdinand-Jühlke-Straße 7
99095 Erfurt, Deutschland
produktsicherheit@kolibri360.de

Druck:
CPI Druckdienstleistungen GmbH
im Auftrag der
Kolibri 360 GmbH
Ferdinand-Jühlke-Str. 7
99095 Erfurt